The Illustrated Flora of Illinois

The Illustrated Flora of Illinois
Robert H. Mohlenbrock, *General Editor*

The Illustrated Flora of Illinois

Flowering Plants

Asteraceae, Part I

Robert H. Mohlenbrock

Illustrated by Paul W. Nelson

Southern
Illinois
University Press

Carbondale

Printed in the United States of America

18 17 16 15 4 3 2 1

Cover and title page illustration: *Solidago flexicaulis* (zigzag goldenrod), by Paul W. Nelson

Library of Congress Cataloging-in-Publication Data
Mohlenbrock, Robert H., 1931–
Flowering plants. Asteraceae. Part 1 / Robert H. Mohlenbrock ; illustrated by Paul W. Nelson.
pages cm—(The illustrated flora of Illinois)
Other title: Asteraceae
Includes bibliographical references and index.
ISBN 978-0-8093-3367-7 (pbk. : alk. paper)
ISBN 0-8093-3367-8 (pbk. : alk. paper)
ISBN 978-0-8093-3368-4 (ebook)
ISBN 0-8093-3368-6 (ebook)
1. Compositae—Illinois. 2. Plants—Illinois. I. Title. II. Title: Asteraceae. III. Series:
Illlustrated flora of Illinois.
QK495.C74M64 2015
583'.9909773—dc23 2014025547

The paper used in this publication meets the minimum requirements of American National Standard for Information Sciences—Permanence of Paper for Printed Library Materials, ANSI Z39.48-1992. ♾

To Henry and Alice Barkhausen,

ardent and dedicated conservationists

and wonderful friends, who have

supported my Illustrated Flora of

Illinois project for many years

Contents

Preface

Several volumes in the Illustrated Flora of Illinois series will be devoted to the dicotyledonous flowering plants; this volume is the eighth one. It follows publication of one on ferns, six on monocotyledonous plants, and seven on other dicots.

The concept of the Illustrated Flora of Illinois is to produce a multivolume flora of the plants of the state of Illinois. For each kind of plant known to occur in Illinois, complete descriptions, illustrations showing diagnostic features, and ecological notes, along with a statement of distribution, are provided.

There is no definite sequence for publication of the Illustrated Flora of Illinois. Volumes will appear as they are completed.

Herbaria from which specimens have been studied are at Eastern Illinois University, the Field Museum of Natural History, the Gray Herbarium of Harvard University, the Illinois Natural History Survey, the Illinois State Museum, the Missouri Botanical Garden, the Morton Arboretum, the New York Botanical Garden, the Shawnee National Forest, Southern Illinois University Carbondale, the United States National Herbarium, the University of Illinois, and Western Illinois University. In addition, some private collections have been examined. The author is indebted to the curators and staffs of these herbaria for the courtesies extended. I am deeply grateful to Henry and Alice Barkhausen, the Gaylord and Dorothy Donnelley Foundation, and Southern Illinois University Press for their generous support that made this volume possible.

The illustrations for each species in this volume, depicting the habit and distinguishing features, were prepared by Paul W. Nelson. My wife, Beverly, assisted me in several of the herbaria and on many of the field trips, and typed drafts of the manuscript. Madison Preece prepared the glossary and the indexes. Without the help of all those individuals and organizations, this book would not have been possible.

The Illustrated Flora of Illinois

Flowering Plants:
Asteraceae, Part 1

County Map of Illinois

Introduction

Flowering plants that form two "seed leaves," or cotyledons, when the seed germinates are called dicotyledons, or dicots. They far exceed the number of species of monocots, or flowering plants that produce a single "seed leaf" upon germination. This is the eighth volume of the Illustrated Flora of Illinois to be devoted to the dicots of Illinois.

The system of classification adopted for the Illustrated Flora of Illinois was proposed by Thorne in 1968. This system is a marked departure from the more familiar system of Engler and Prantl. This latter system, which is still followed in many regional floras, is out-of-date and does not reflect the vast information recently gained from the study of cytology, biochemistry, anatomy, and embryology. In fact, the Thorne system no longer depicts many of the relationships adhered to in 1968.

Since the arrangement of orders and families proposed by Thorne is unfamiliar to many, an outline of the orders and families of flowering plants known to occur in Illinois is presented.

Those names in boldface are orders and families for which books have already been published in this series. The family in all capital letters is the one described in this volume of the Illustrated Flora of Illinois.

Order Annonales
Family Magnoliaceae
Family Annonaceae
Family Calycanthaceae
Family Aristolochiaceae
Family Lauraceae
Family Saururaceae

Order Berberidales
Family Menispermaceae
Family Ranunculaceae
Family Berberidaceae
Family Papaveraceae

Order Nymphaeales
Family Nymphaeaceae
Family Ceratophyllaceae

Order Sarraceniales
Family Sarraceniaceae

Order Theales
Family Aquifoliaceae
Family Hypericaceae
Family Elatinaceae
Family Ericaceae

Order Ebenales
Family Ebenaceae
Family Styracaceae
Family Sapotaceae

Order Primulales
Family Primulaceae

Order Cistales
Family Violaceae
Family Cistaceae
Family Passifloraceae
Family Cucurbitaceae
Family Loasaceae

Order Salicales
Family Salicaceae

Order Tamaricales
Family Tamaricaceae

Order Capparidales
Family Capparidaceae
Family Resedaceae
Family Brassicaceae

Order Malvales
Family Sterculiaceae
Family Tiliaceae
Family Malvaceae

Order Urticales
Family Ulmaceae
Family Moraceae
Family Urticaceae

Order Rhamnales
Family Rhamnaceae
Family Elaeagnaceae

Order Euphorbiales
Family Thymelaeaceae
Family Euphorbiaceae

Order Solanales
Family Solanaceae
Family Convolvulaceae
Family Cuscutaceae
Family Polemoniaceae

Order Campanulales
Family Campanulaceae

Order Santalales
Family Celastraceae
Family Santalaceae
Family Loranthaceae

Order Oleales
Family Oleaceae

Order Geraniales
Family Linaceae
Family Zygophyllaceae
Family Oxalidaceae
Family Geraniaceae
Family Balsaminaceae
Family Limnanthaceae
Family Polygalaceae

Order Rutales
Family Rutaceae
Family Simaroubaceae
Family Anacardiaceae
Family Sapindaceae
Family Aceraceae
Family Hippocastanaceae
Family Juglandaceae

Order Myricales
Family Myricaceae

Order Chenopodiales
Family Phytolaccaceae
Family Nyctaginaceae
Family Aizoaceae
Family Cactaceae
Family Portulacaceae
Family Chenopodiaceae
Family Amaranthaceae
Family Caryophyllaceae
Family Polygonaceae

Order Hamamelidales
Family Hamamelidaceae
Family Platanaceae

Order Fagales
Family Fagaceae
Family Betulaceae
Family Corylaceae

Order Rosales
Family Rosaceae
Family Mimosaceae
Family Caesalpiniaceae
Family Fabaceae
Family Crassulaceae
Family Penthoraceae
Family Saxifragaceae
Family Droseraceae
Family Staphyleaceae

Order Myrtales
Family Lythraceae
Family Melastomaceae
Family Onagraceae

Order Gentianales
Family Loganiaceae
Family Rubiaceae
Family Apocynaceae
Family Asclepiadaceae
Family Gentianaceae
Family Menyanthaceae

Order Bignoniales
Family Bignoniaceae
Family Martyniaceae
Family Scrophulariaceae
Family Paulowniaceae
Family Plantaginaceae
Family Orobanchaceae
Family Lentibulariaceae
Family Acanthaceae

Order Cornales
Family Vitaceae
Family Nyssaceae
Family Cornaceae
Family Haloragidaceae
Family Hippuridaceae
Family Araliaceae
Family Apiaceae

Order Dipsacales
Family Caprifoliaceae
Family Adoxaceae
Family Valerianaceae
Family Dipsacaceae

Order Lamiales
Family Hydrophyllaceae
Family Heliotropaceae
Family Boraginaceae
Family Verbenaceae
Family Phrymaceae
Family Callitrichaceae
Family Lamiaceae

ORDER ASTERALES
FAMILY ASTERACEAE

Since only one part of one family is included in this book, no general key to the dicot families has been provided. The reader is invited to use my companion book *Guide to the Vascular Flora of Illinois* (2013) for keys to all families of flowering plants in Illinois.

The nomenclature for the species and lesser taxa used in this volume has been arrived at after lengthy study of recent floras and monographs. Synonyms, with complete author citations, that have been applied to species and lesser taxa in Illinois are given under species names. A description, while not necessarily intended to be complete, covers the important features of the species.

The common name, or names, is the one used locally in Illinois. The habitat designation is not always the habitat throughout the range of the species, but only for it in Illinois. The overall range for each species is given from the northeastern to the northwestern extremities, south to the southwestern limit, then eastward to the southeastern limit. The range has been compiled from various sources, including examination of herbarium material and field studies of my own. A general statement is given concerning the range of each species in Illinois.

Descriptions and Illustrations

This is the first of three volumes describing the family Asteraceae in Illinois.

This family for many years was called the Compositae. Some botanists in the past, with sound reasoning, divided the family into three families, but I have chosen to follow the current belief that these three segregates should be treated as a single but diverse family.

In Illinois, I am recognizing 388 species in 119 genera, as well as 31 hybrids and 73 lesser taxa, making Asteraceae the largest family of flowering plants in the state. Of these 388 species, 127 of them, or roughly 33%, are non-natives. The Poaceae, or grass family, the second largest in Illinois, has 353 species in 101 genera, along with 5 hybrids and 36 lesser taxa.

Worldwide, members of the Asteraceae exhibit nearly every growth form, but in Illinois, all species are herbaceous and include annuals, biennials, and perennials. Only one species, *Mikania scandens*, is a vine. All the rest are either erect, ascending, or prostrate herbs. Some species have latex. Many plants have taproots or fibrous roots, or both, although there are a number of rhizomatous or stoloniferous species.

With such a large and diverse family, the species exhibit every type of leaf possible. A few species have only basal leaves; others have only cauline leaves; still others have both basal and cauline leaves. In this last group, the basal leaves may or may not be present at flowering time. In those plants with cauline leaves, the leaves may be alternate, opposite, or whorled. Leaves may be simple, unlobed or lobed, or variously compound. Nearly every type of pubescence may be found in the family.

Flowers occur in heads, known as capitulae, with each head having few to numerous flowers. The heads are usually several and are arranged in a variety of inflorescences, called arrays, that may be in the form of corymbs, cymes, panicles, thyrses, or racemes. Occasionally the head may be reduced to one per plant. Each head is surrounded by one or more series of bracts, known as phyllaries. The phyllaries make up the involucre. In a given head, the phyllaries may be equal or unequal. They may be green and herbaceous, or they may have a scarious margin. They may be glabrous or variously pubescent and glandular or eglandular. In *Coreopsis* and *Taraxacum*, there are tiny bractlets, called calyculi, at the outside base of the phyllaries.

Heads may have ray flowers or disc flowers, or both. Ray flowers have rays that are zygomorphic and often erose or shallowly lobed at the tip. They may have 5 stamens and an inferior ovary. Rays are sometimes referred to as petals by the uninformed. Disc flowers are actinomorphic, usually short-tubular, with 4 or usually 5 lobes.

Experts in this family recognize different types of heads. Radiate heads have peripheral rays that may be pistillate or sterile, as well as central disc flowers that are either bisexual or staminate. Liguliferous heads have only ray flowers that are bisexual. Discoid heads have only disc flowers that may be bisexual, only pistillate, or only staminate. In disciform heads, all flowers are disc flowers, but the peripheral flowers have filiform corollas that are usually only pistillate.

All flowers in a head share a common receptacle. The receptacle may be flat or convex. It may bear tiny scales called paleae. The paleae of a receptacle are referred to as chaff. Receptacles without paleae are said to be epaleate or naked. When the paleae are shed, they may leave either a smooth or a pitted receptacle. Occasionally, hairs, scales, or bristles may also be present on the receptacle.

Although the fruits of the Asteraceae are often referred to as achenes, they are actually cypselae. Achenes are dry, one-seeded fruits that are derived from flowers with a unicarpellate superior ovary. Cypselae are dry, one-seeded fruits that are derived from flowers with a bicarpellate inferior ovary. The cypselae may be crowned or subtended by pappi, which may be in the form of capillary or plumose bristles, awns, or scales. The pappus is usually thought to be the remains of a calyx. In a few species, the pappi are absent.

Although molecular phylogenetic studies within the Asteraceae have resulted in different classifications within the family, I am following the more traditional divisions of the family into tribes used in *Flora of North America.*

Below are listed the tribes of Asteraceae in North America, along with the number of species in North America and the number of species in Illinois.

Tribe Mutisieae	7 genera, 14 species in U.S.—None in Illinois
Tribe Cynareae	17 genera, 116 species in U.S.
	10 genera, 34 species, 1 hybrid in Illinois
Tribe Arctotideae	3 genera, 4 species in U.S.—None in Illinois
Tribe Vernonieae	6 genera, 25 species in U.S.
	2 genera, 6 species, 1 hybrid in Illinois
Tribe Cichorieae	49 genera, 229 species in U.S.
	19 genera, 50 species, 2 hybrids in Illinois
Tribe Calenduleae	4 genera, 7 species in U.S.—None in Illinois
Tribe Gnaphalieae	19 genera, 101 species in U.S.
	5 genera, 10 species in Illinois
Tribe Inuleae	3 genera, 5 species in U.S.
	1 genus, 1 species in Illinois
Tribe Senecioneae	29 genera, 167 species in U.S.
	8 genera, 17 species in Illinois
Tribe Plucheae	3 genera, 12 species in U.S.
	1 genus, 2 species in Illinois
Tribe Anthemideae	26 genera, 99 species in U.S.
	9 genera, 24 species in Illinois
Tribe Astereae	77 genera, 719 species in U.S.
	19 genera, 103 species, 2 hybrids in Illinois
Tribe Heliantheae	148 genera, 746 species in U.S.
	42 genera, 139 species, 20 hybrids in Illinois
Tribe Eupatorieae	27 genera, 159 species in U.S.
	8 genera, 25 species, 5 hybrids in Illinois

Because of the large number of taxa of Asteraceae in Illinois, the treatment of them will be in three volumes. Since goldenrods and asters are the largest groups and often are confusing, I have elected to treat them, along with other members of the tribe Astereae, in the initial volume. However, to have each of the three volumes more or less comparable in size, I have added the tribe Anthemideae to this book.

I do not want to divide any tribe into more than one volume, so the tribes Heliantheae and Senecioneae will comprise the second volume, although it will treat a few more taxa than the other volumes.

The third book in this series encompasses the other tribes in the Asteraceae: Cynareae, Vernonieae, Gnaphalieae, Inuleae, Eupatorieae, Plucheae, and Cichorieae.

After the description of the Asteraceae and the key to all the genera in Illinois, the description, habitat notes, nomenclatural issues, uses, and other applicable information will be provided for each taxon. Following the name of each taxon are any synonyms that may be pertinent.

Order Asterales

This order consists only of the family Asteraceae.

Family Asteraceae—Aster Family

Annual, biennial, or perennial herbs (in Illinois), rarely a vine (*Mikania scandens*), usually with a taproot, but sometimes rhizomatous or stoloniferous; latex present in some species; leaves basal or cauline, or both, the cauline ones alternate, opposite, or whorled; flowers 2 to numerous in heads, the heads arranged in inflorescences called arrays, sometimes in corymbs, cymes, panicles, or racemes, rarely the head solitary; heads subtended by 1 to several bracts, called phyllaries, the phyllaries forming an involucre; heads with ray flowers, disc flowers, or both; ray flowers consisting of a flat lamina of various colors, some of them bisexual, some unisexual, or some neutral, often notched or erose at the tip, sometimes with 5 stamens and an inferior ovary; disc flowers tubular, some of them bisexual, some unisexual, or some neutral, usually 5-lobed (rarely 4-lobed), sometimes with 5 stamens and an inferior ovary; receptacle flat or convex, with or without scales or chaff, known as paleae, that subtend each flower; fruit a cypsela, sometimes crowned by a pappus of capillary bristles, plumose bristles, scales, or awns, or pappus absent.

Worldwide the family Asteraceae consists of about 23,000 species in about 1,500 genera.

Key to the Genera of Asteraceae in Illinois

Names appearing in boldface are treated in this book; if the genus is in the second volume (*Asteraceae, Part 2*), the name is followed by *II*; if the genus is in the third volume (*Asteraceae, Part 3*), the name is followed by *III*.

1. Flowering heads with only ray flowers; latex present .... Group 1
1. Flowering heads with disc flowers, the ray flowers present or absent; latex absent.
   2. All the leaves, or at least those on the lower part of the stem, opposite or whorled .... Group 2

2. None of the leaves opposite or whorled.
3. Ray flowers present.
4. Rays yellow or orange .... Group 3
4. Rays blue, purple, pink, rose, or white, not yellow nor orange .... Group 4
3. Ray flowers absent.
5. Leaves simple, entire, toothed, or shallowly lobed .... Group 5
5. Leaves deeply pinnatifid or pinnately compound .... Group 6

Group 1

Flowering heads with only ray flowers; latex present.

1. Flowers blue.
2. Heads at least 2.5 cm across; cypselae not flat; phyllaries in 2 series; pappus of 2 or 3 series of reduced scales .... 100. *Cichorium* III
2. Heads less than 2.5 cm across; cypselae flat; phyllaries in 3 or 4 series; pappus of capillary bristles.
3. Cypselae beaked .... 106. *Lactuca* III
3. Cypselae not beaked .... 105. *Mulgedium* III
1. Flowers yellow, orange, purple, pink, cream, or white.
4. Flowers pink, cream, or white .... 107. *Nabalus* III
4. Flowers yellow, orange, or purple.
5. Cypselae flat.
6. Teeth of leaves usually spine-tipped; phyllaries in several series; cypselae without a beak; flowers 80 or more per head .... 108. *Sonchus* III
6. Teeth of leaves rarely spine-tipped; phyllaries in 3 or 4 series; cypselae beaked; flowers up to 50 per head .... 106. *Lactuca* III
5. Cypselae terete or angular, not flat.
7. Pappus absent, or of 1–3 small bristles.
8. Phyllaries numerous, in several series; flowers usually orange ... 117. *Serinia* III
8. Phyllaries 8 in a single series; flowers usually yellow .... 104. *Lapsana* III
7. Pappus present, consisting of several bristles.
9. Pappus only of capillary bristles.
10. Cauline leaves present.
11. None of the leaves pinnatifid; phyllaries in several series .... 109. *Hieracium* III
11. Some or all of the leaves pinnatifid; phyllaries often in one or two series.
12. Bractlets 5–12; ray flowers 15–30; cypselae not beaked .... 101. *Crepis* III
12. Bractlets 3–4; ray flowers 7–15; cypselae beaked ... 103. *Chondrilla* III
10. Leaves all, or nearly all, basal.
13. Cypselae beaked.
14. Pappus white; phyllaries in 2 series .... 102. *Taraxacum* III
14. Pappus sordid; phyllaries in several series .... 118. *Pyrrhopappus* III
13. Cypselae not beaked.
15. Head solitary .... 115. *Nothocalais* III
15. Heads several .... 101. *Crepis* III
9. Pappus of plumose bristles, or with both scales and bristles.
16. Pappus with scales and bristles .... 116. *Krigia* III

16. Pappus of plumose bristles.
  17. Phyllaries in several series; leaves mostly basal.
    18. Receptacle paleate .... 111. *Hypochaeris* III
    18. Receptacle villous but not paleate .... 110. *Leontodon* III
  17. Phyllaries in 1 series; leaves cauline.
    19. Stems spinescent.
      20. Outer phyllaries ovate, 3.5–8.0 mm wide; cypselae with a distinct slender beak .... 113. *Picris* III
      20. Outer phyllaries linear-lanceolate, less than 3 mm wide; cypselae with little or no beak .... 112. *Helminotheca* III
    19. Stems not spinescent .... 114. *Tragopogon* III

Group 2

Flowering heads with disc flowers, the ray flowers present or absent; all the leaves, or at least those on the lower part of the stem, opposite or whorled; latex absent.

1. Ray flowers present.
  2. Rays yellow or orange.
    3. Pappus of numerous capillary bristles; plants creeping ...... 51. *Calyptocarpus* II
    3. Pappus of awns, scales, 1–3 small bristles, or absent; plants upright (procumbent in *Sanvitalia*).
      4. Leaves simple, entire, serrate, shallowly lobed, or palmately lobed.
        5. Leaves shallowly palmately lobed; phyllaries in 1 series ... 35. *Smallanthus* II
        5. Leaves not palmately lobed; phyllaries in 2 or more series.
          6. Disc flowers sterile with poorly developed cypselae .... 43. *Silphium* II
          6. Disc flowers fertile with well-developed cypselae.
            7. Ray flowers persistent on the cypselae................. 41. *Heliopsis* II
            7. Ray flowers deciduous from the cypselae.
              8. Pappus absent or of 1–3 small bristles.
                9. Petioles absent or nearly so; cypselae 4–5 mm long, strongly angular; pappus completely absent................37. *Guizotia* II
                9. Petioles at least 3 mm long; cypselae 1.0–2.5 mm long, weakly angular; pappus absent or of 1–3 small bristles .... 50. *Acmella* II
              8. Pappus of awns or scales, rarely absent.
                10. Cypselae flat.
                  11. Stems winged.
                    12. Cypselae wingless; all leaves opposite......45. *Verbesina* II
                    12. Cypselae winged; some of the leaves alternate .... 46. *Actinomeris* II
                  11. Stems wingless ....47. *Ximenesia* II
                10. Cypselae angular.
                  13. Cypselae 3-angled; pappus of cypselae of ray flowers with 3 awns .... 42. *Sanvitalia* II
                  13. Cypselae 2- or 4-angled; pappus of cypselae of ray flowers with 2 or 4 awns.
                    14. Cypselae with 2 or 4 stout, barbed awns ..... 57. *Bidens* II
                    14. Cypselae with 2 small, barbless awns, or pappus rarely absent.

15. Rays usually 8 per head; phyllaries in 2 series ....
.................................54. *Coreopsis* II
15. Rays of various numbers per head, not all of them 8; phyllaries in several series.
16. Cypselae wingless.
17. Ray flowers neutral ....52. *Helianthus* II
17. Ray flowers fertile ....45. *Verbesina* II
16. Cypselae winged................47. *Ximenesia* II
4. Leaves deeply pinnately lobed, or 1- to 2-pinnate.
18. Plants aquatic; submerged leaves different from emergent leaves..........
....58. *Megalodonta* II
18. Plants terrestrial; submerged leaves absent.
19. Pappus of 20 scales, each with 10 bristles at tip; receptacle bristly .....
.... 59. *Dyssodia* II
19. Pappus of small scales, not bristle-tipped, with 2 or 4 awns; receptacle paleate.
20. Phyllaries in 1 series; pappus of unequal scales ....60. *Tagetes* II
20. Phyllaries in 2 series; pappus of 2–8 awns.
21. Leaves once-divided.
22. Pappus of 2 or 4 stout, barbed awns .... *Bidens* II
22. Pappus of 2 weak, barbless awns ....54. *Coreopsis* II
21. Leaves at least twice-divided .... 56. *Cosmos* II
2. Rays white, pink, or rose.
23. Leaves pinnately divided or simple and lobed.
24. Some of the leaves pinnately divided; phyllaries in 2 series; pappus present.
25. Leaves and leaflets more than 1 cm wide .... 57. *Bidens* II
25. Leaflets 1–2 mm wide................................ 56. *Cosmos* II
24. Leaves simple and lobed; phyllaries in 2 series; pappus absent.............
....36. *Polymnia* II
23. Leaves simple, unlobed.
26. Pappus of ray flowers consisting of 15–20 fimbriate scales.....53. *Galinsoga* II
26. Pappus of ray flowers reduced to a small crown or absent.
27. Plants more than 50 cm tall; stems green...............45. *Verbesina* II
27. Plants usually less than 50 cm tall; stems mauve ....49. *Eclipta* II
1. Ray flowers absent.
28. Flowers unisexual, green.
29. Pistillate involucres nutlike or burlike ....30. *Ambrosia* II
29. Pistillate involucres not nutlike nor burlike.
30. Heads in racemose spikes, bracteate .... 33. *Iva* II
30. Heads in paniculate spikes, ebracteate .... 34. *Cyclachaena* II
28. Flowers perfect, not green.
31. Plants climbing; major phyllaries 4 per head, subtended by short outer phyllaries ....79. *Mikania* III
31. Plants erect; phyllaries not as above.
32. Some of the leaves divided into 3–7 leaflets or 2- to 3-pinnate.
33. Leaves once-pinnate................................ 57. *Bidens* II
33. Leaves 2- to 3-pinnate .... 55. *Thelesperma* II
32. Leaves simple, sometimes lobed but not compound.
34. Pappus of 2 or 4 stout, barbed awns.

35. Pappus of 2 or 4 stout, barbed awns; disc yellow . . . . . . 57. *Bidens* II
35. Pappus of aristate scales; disc blue or purple. . . . . . . 75. *Ageratum* III
34. Pappus of capillary bristles, or pappus absent.
36. Pappus absent; receptacle paleate; stems square . . . . 48. *Melanthera* II
36. Pappus of capillary bristles; receptacle naked; stems not square.
37. Some or all the leaves whorled; flowers purple or rose . . . . 73. *Eutrochium* III
37. Leaves opposite (if rarely whorled, the flowers white); flowers white, blue, or pink.
38. Receptacle conical; flowers blue . . . . 74. *Conoclinium* III
38. Receptacle flat; flowers white or pink.
39. Flowers pink . . . . 78. *Fleischmannia* III
39. Flowers white.
40. Phyllaries all of same length . . . . 80. *Ageratina* III
40. Phyllaries of different lengths . . . . 72. *Eupatorium* III

Group 3

Plants with both ray and disc flowers; rays yellow or orange; leaves alternate or basal; latex absent.

1. Leaves simple, entire, serrate, or shallowly lobed.
2. Most or all the leaves basal.
3. Head solitary.
4. Pappus of 5–8 translucent scales . . . . 64. *Tetraneuris* II
4. Pappus of capillary bristles . . . . 70. *Tussilago* II
3. Heads several . . . . 66. *Packera* II
2. Most of the leaves cauline.
5. Rays reflexed; disc long-columnar or globose.
6. Disc long-columnar; cauline leaves clasping . . . . 38. *Dracopis* II
6. Disc globose; cauline leaves not clasping . . . . 63. *Helenium* II
5. Rays spreading, rarely slightly reflexed; disc globose, conical, or flat.
7. Pappus entirely of capillary or barbellate bristles.
8. Leaves spinulose-dentate . . . . 17. ***Prionopsis***, p. 130
8. Leaves not spinulose-dentate.
9. Upper leaves clasping . . . . 86. *Inula* III
9. None of the leaves clasping.
10. Inflorescence more or less a flat-topped corymb.
11. Leaves up to 5 (–12) mm wide, some of them usually glandular-punctate . . . . 7. ***Euthamia***, p. 31
11. Leaves usually more than 5 mm wide, not glandular-punctate. . . . . 9. ***Oligoneuron***, p. 92
10. Inflorescence paniculate, thyrsoid, or in axillary clusters . . . . 8. ***Solidago***, p. 38
7. Pappus of a short crown, awns, or scales or, if capillary bristles present, scales are also present.
12. Receptacle bristly; pappus of 6–10 awned scales . . . . 62. *Gaillardia* II
12. Receptacle epaleate or paleate; pappus not as above.
13. Pappus of the cypselae of the disc flowers scaly, with short teeth, a crown, or absent.

14. Outer pappus scaly, inner pappus bristly on the cypselae of the disc flowers.
15. Cypselae of ray flowers thick, of disc flowers flat .... 11. ***Heterotheca***, p. 103
15. All cypselae flat .... 10. ***Chrysopsis***, p. 100
14. Pappus of the cypselae of the disc flowers with short teeth, a crown, or absent.
16. Disc flowers sterile; phyllaries in 2 or 3 series........43. *Silphium* II
16. Disc flowers fertile; phyllaries in several series...... 39. *Rudbeckia* II
13. Pappus of the cypselae of the disc flowers with 2–8 awns.
17. Phyllaries gummy......................... 18. ***Grindelia***, p. 132
17. Phyllaries not gummy.
18. Pappus of 2 awns, these sometimes deciduous.
19. Disc flowers sterile; phyllaries in 2 or 3 series...43. *Silphium* II
19. Disc flowers perfect; phyllaries in several series.
20. Cypselae wingless ....52. *Helianthus* II
20. Cypselae winged.
21. Stems unwinged...................47. *Ximenesia* II
21. Stems winged .... 46. *Actinomeris* II
18. Pappus of 5 or more awns, persistent.
22. Disc flowers sterile .... 5. ***Amphiachyris***, p. 27
22. Disc flowers fertile .... 6. ***Gutierrezia***, p. 29
1. Leaves simple and deeply divided, or leaves compound.
23. Receptacle epaleate.
24. Rays reflexed; pappus reduced to a short crown, or absent; phyllaries in 2 series ....40. *Ratibida* II
24. Rays spreading; pappus of barbellate bristles; phyllaries in 1 series.
25. Stems glandular-pubescent; rays 1–3 mm long ....65. *Senecio* II
25. Stems eglandular; rays 6–10 mm long.
26. Leaves once-pinnatifid .... 66. *Packera* II
26. Leaves 2- to 3-pinnatifid..............................65. *Senecio* II
23. Receptacle paleate.
27. Receptacle flat; phyllaries in 2 or 3 series; disc flowers sterile.... 43. *Silphium* II
27. Receptacle conical; phyllaries in several series; disc flowers fertile.
28. Leaves 2- to 3-pinnate or -pinnatifid ....27. ***Cota***, p. 236
28. Leaves lobed or 1-pinnatifid............................ 39. *Rudbeckia* II

Group 4

Plants with both ray flowers and disc flowers; rays not orange nor yellow; latex absent; leaves alternate or basal.

1. Leaves simple and entire, serrate, or shallowly lobed (sometimes with a basal pair of pinnae in *Tanacetum*).
2. Rays blue, purple, or pink.
3. Rays reflexed; disc conical .... 44. *Echinacea* II
3. Rays spreading; disc flat or subglobose.
4. Pappus a double series of capillary bristles .... 4. ***Ionactis***, p. 25
4. Pappus a single series of capillary bristles, awns, scales, or absent.
5. Leaves mostly basal; flower heads 1 to a few .... 2. ***Bellis***, p. 21

5. Leaves mostly cauline; flower heads usually numerous.
6. Pappus of capillary bristles.
7. Rays usually more than 50 . . . . 12. ***Erigeron***, p. 105
7. Rays usually fewer than 50.
8. Basal leaves cordate; inflorescence a flat-topped corymb . . . . 15. ***Eurybia***, p. 122
8. Basal leaves tapering or rounded at the base or, if cordate, the inflorescence racemose or paniculate.
9. Basal leaves cordate . . . . 19. ***Symphyotrichum***, p. 137
9. Basal leaves tapering or rounded at the base, or absent at flowering time.
10. Annuals . . . . 19. ***Symphyotrichum***, p. 137
10. Perennials.
11. Leaves sericeous on both surfaces . . . . 19. ***Symphyotrichum***, p. 137
11. Leaves glabrous or pubescent, but not sericeous.
12. Stems glandular, at least above; leaves oblong; phyllaries densely glandular . . . . 19. ***Symphyotrichum***, p. 137
12. Stems eglandular; leaves linear or lanceolate or elliptic; phyllaries eglandular.
13. Leaves coarsely toothed, all of them 2.5 cm wide or wider . . . . 1. ***Aster***, p. 19
13. Leaves entire or sparsely and finely toothed, many of them less than 2.5 cm wide. . . . . 19. ***Symphyotrichum***, p. 137
6. Pappus of short awns or scales.
14. Pappus of 6–10 awned scales; rays purple; phyllaries in 2 or 3 series; receptacle setose . . . . 62. *Gaillardia* II
14. Pappus of 2 awns and tiny bristles; rays pink; phyllaries in several series; receptacle epaleate . . . . 14. ***Boltonia***, p. 116
2. Rays white.
15. Pappus of capillary bristles.
16. Pappus in a double series.
17. Inflorescence flat-topped . . . . 3. ***Doellingeria***, p. 22
17. Inflorescence paniculate . . . . 4. ***Ionactis***, p. 25
16. Pappus in a single series.
18. Rays less than 5 mm long . . . . 13. ***Conyza***, p. 113
18. Rays more than 5 mm long.
19. Inflorescence flat-topped . . . . 9. ***Oligoneuron***, p. 92
19. Inflorescence paniculate, not flat-topped.
20. Rays usually more than 50. . . . . 12. ***Erigeron***, p. 105
20. Rays usually fewer than 50 . . . . 19. ***Symphyotrichum***, p. 137
15. Pappus of 2 or 3 awns, a short crown, or absent.
21. Pappus absent; cypselae 3-angled. . . . . 23. ***Chamaemelum***, p. 213
21. Pappus present; cypselae not 3-angled.
22. Pappus reduced to a short crown; leaves sometimes with a basal pair of pinnae. . . . . 21. ***Tanacetum***, p. 206
22. Pappus of 2 or 3 awns; leaves without a basal pair of pinnae.
23. Stems winged . . . . 45. *Verbesina* II

23. Stems unwinged.
24. Pappus of 2 awns and tiny bristles; plants glabrous; receptacle naked . . . . 14. ***Boltonia***, p. 116
24. Pappus of 2 or 3 awns, without bristles; plants pubescent; receptacle paleate . . . . 32. *Parthenium* II

1. Leaves deeply pinnatifid or 1- to 3-pinnate.
25. Leaves merely deeply lobed or pinnatisect.
26. Pappus absent; receptacle naked . . . . 29. ***Leucanthemum***, p. 240
26. Pappus or 2 or 3 awns; receptacle paleate. . . . . 32. *Parthenium* II
25. Leaves 1- to 3-pinnate.
27. Pappus of capillary bristles; phyllaries more or less squarrose; flowers blue or purple . . . . 16. ***Machaeranthera***, p. 128
27. Pappus a low crown, or absent; phyllaries not squarrose; flowers white.
28. Receptacle paleate, phyllaries in several series; pappus absent.
29. Plants aromatic; receptacle conical. . . . . 25. ***Anthemis***, p. 231
29. Plants not aromatic; receptacle flat. . . . . 22. ***Achillea***, p. 211
28. Receptacle epaleate; phyllaries in 2 or 3 series; pappus a low crown, or absent.
30. Cypselae 3-ribbed; plants not aromatic . . . . .28. ***Tripleurospermum***, p. 238
30. Cypselae 5-ribbed; plants aromatic. . . . . .26. ***Matricaria***, p. 234

Group 5

Flowering heads with only disc flowers; leaves alternate or basal and simple, entire, serrate, shallowly lobed, or pinnatifid (or with a basal pair of pinnae in *Tanacetum*); latex absent.

1. Leaves with spine-tipped teeth.
2. Flowers yellow . . . . 97. *Centaurea* III
2. Flowers purple, pink, pale blue, or white.
3. Pappus of barbellate bristles; phyllaries in 2 series; stems winged. . . . . 89. *Onopordum* III
3. Pappus of simple bristles, plumose bristles, or a crown of scales; phyllaries in several series; stems unwinged.
4. Pappus a crown of scales; flowers pale blue . . . .*Echinops* III
4. Pappus of bristles; flowers purple, white, or pink.
5. Pappus of plumose bristles. . . . . 91. *Cirsium* III
5. Pappus of simple bristles . . . . .90. *Carduus* III

1. Leaves without spine-tipped teeth.
6. Outer row of disc flowers appearing ligulate.
7. Outer phyllaries entire, never spine-tipped.
8. Phyllaries coriaceous, yellowish, without a hyaline margin . . . . 94. *Amberboa* III
8. Phyllaries thin, green, with a hyaline margin . . . . 93. *Acroptilonn* III
7. Outer phyllaries fimbriate or laciniate, some of them spine-tipped.
9. Involucre 3–4 cm high; pappus 6–12 mm long . . . . .95. *Plectocephalus* III
9. Involucre 1.0–1.5 mm high; pappus 3 mm long or less . . . . 97. *Centaurea* III
6. Outer row of disc flowers tubular, not appearing ligulate.
10. Flowers greenish.
11. Phyllaries and fruits with hooked bristles, prickly or appearing prickly.
12. Flowers unisexual . . . . 31. *Xanthium* II
12. Flowers perfect. . . . . .92. *Arctium* III

11. Phyllaries and fruits without hooked bristles, not prickly .... 24. ***Artemisia***, p. 215

10. Flowers white, purple, cream, yellow, orange, blue, pink, brownish, or rusty.

13. Phyllaries and fruits with hooked bristles .... 92. *Arctium* III

13. Phyllaries and fruits without hooked bristles.

14. Flowers orange; pappus absent or of short scales; some leaves clasping; receptacle paleate .... 96. *Carthamus* III

14. Flowers not orange; pappus of capillary bristles, a short crown, or 5–8 scales; leaves not clasping; receptacle epaleate.

15. Flowers yellow, cream, brownish, or rusty.

16. Phyllaries in 1 series; flowers yellow .... 65. *Senecio* II

16. Phyllaries in several series; flowers cream, brownish, or rusty.

17. Pappus of plumose bristles .... 76. *Brickellia* III

17. Pappus of capillary bristles.

18. Heads leafy bracted .... 84. *Gnaphalium* III

18. Heads not leafy bracted .... 82. *Pseudognaphalium* III

15. Flowers pink, purple, or white.

19. Flowers pink or purple.

20. Flowers in glomerules, subtended by a 3-lobed bract.. 98. *Elephantopus* III

20. Flowers and bracts not as above.

21. Phyllaries in 1 series, subtended by calyxlike bracts; leaves basal.................................. 71. *Petasites* II

21. Phyllaries in several series; leaves not basal.

22. Pappus of plumose or barbellate bristles.. 77. *Liatris* III

22. Pappus of simple bristles.

23. Pappus in a double series .... 99. *Vernonia* III

23. Pappus in a single series.

24. Capillary bristles united at base.. 85. *Gamochaeta* III

24. Capillary bristles not united at base .... 87. *Pluchea* III

19. Flowers white.

25. Phyllaries in 1 series.

26. Calyxlike bracts at base of phyllaries; leaves basal.. 71. *Petasites* II

26. Calyxlike bracts absent; leaves cauline.

27. Phyllaries 5; flowers 5 per head .... 69. *Arnoglossum* II

27. Phyllaries more than 5; flowers more than 5 per head .... 67. *Erechtites* II

25. Phyllaries in 2 to several series.

28. Some of the leaves hastate............... 68. *Hasteola* II

28. None of the leaves hastate.

29. Pappus a short crown .... 21. ***Tanacetum***, p. 206

29. Pappus of capillary bristles.

30. Phyllaries in 2 series, not scarious.. 20. ***Brachyactis***, p. 203

30. Phyllaries in several series, scarious.

31. Most of the leaves near the base of the plant .. ..81. *Antennaria* III
31. Most of the leaves cauline .....83. *Anaphalis* III

Group 6

Flowering heads with only disc flowers; leaves alternate or basal, pinnatifid to pinnately compound; latex absent.

1. Flowers greenish.
  2. Pistillate involucre with 2 flowers and 2 beaks; receptacle paleate.... 30. *Ambrosia* II
  2. Involucres not as above; receptacle epaleate or woolly .... 24. ***Artemisia***, p. 215
1. Flowers yellow or greenish yellow.
  3. Pappus of capillary bristles; phyllaries in 1 series ....65. *Senecio* II
  3. Pappus of 12–20 hyaline scales, of short awns, or absent; phyllaries in 2–4 series.
    4. Pappus of 12–20 hyaline scales..........................61. *Hymenopappus* II
    4. Pappus a short crown, or absent.
      5. Receptacle conic; plants aromatic.................... 26. ***Matricaria***, p. 234
      5. Receptacle flat; plants not aromatic .... 21. ***Tanacetum***, p. 206

Tribe Astereae

Annual or perennial herbs (in Illinois); leaves basal or cauline or both, alternate, simple, sometimes deeply lobed, or compound; flowers numerous, usually in 1 to many radiate heads; involucre hemispheric to obconic to campanulate; phyllaries several, equal or unequal, in 1–5 series, appressed to spreading to reflexed; receptacle flat to convex, or with ridges, sometimes pitted, without paleae; ray flowers several, yellow, white, purple, blue, or pink, rarely absent, pistillate; disc flowers usually numerous, usually yellow, the corolla 4- or 5-lobed, perfect; pappus in 1 or 2 series, mostly of capillary bristles thickened at base or uniform throughout or rarely of scales.

In Illinois, this tribe consists of 19 genera, 103 species, 2 hybrids, and 17 lesser taxa.

Key to the Genera of Astereae in Illinois

1. Rays minute or absent, shorter than the disc flowers................ 20. *Brachyactis*
1. Rays present, longer than the disc flowers.
  2. Ray flowers yellow or orange.
    3. Pappus entirely of capillary bristles.
      4. Leaves spinulose-dentate ....17. *Prionopsis*
      4. Leaves not spinulose-dentate.
        5. Inflorescence more or less a flat-topped or round-topped corymb.
          6. Leaves up to 5 mm wide, with parallel veins, some of the leaves glandular-punctate .... 7. *Euthamia*
          6. Leaves usually more than 5 mm wide, without parallel veins, none of the leaves glandular-punctate .... 9. *Oligoneuron*
        5. Inflorescence paniculate, thyrsoid, or in axillary clusters.........8. *Solidago*
    3. Pappus of a short crown, awns, or scales or, if capillary bristles present, scales are present in addition.
      7. Outer pappus scaly, inner pappus bristly on the cypselae of the disc flowers.
        8. Cypselae of ray flowers thick, of disc flowers flat ....11. *Heterotheca*

8. All cypselae flat . . . . 10. *Chrysopsis*
7. Pappus of the disc flowers with 2–8 awns.
9. Phyllaries gummy . . . . 18. *Grindelia*
9. Phyllaries not gummy.
10. Disc flowers sterile. . . . . 5. *Amphyachyris*
10. Disc flowers fertile . . . . 6. *Gutierrezia*
2. Ray flowers white, cream, pink, blue, or purple
11. Rays blue, purple, or pink.
12. Leaves pinnatifid or deeply lobed . . . . 16. *Machaeranthera*
12. Leaves neither pinnatifid nor deeply lobed.
13. Pappus a double series of capillary bristles . . . . 4. *Ionactis*
13. Pappus a single series or 3 or 4 series of capillary bristles, awns, scales, or absent.
14. Leaves all basal; flower heads 1 to a few . . . . 2. *Bellis*
14. Leaves both basal and cauline; heads several to many.
15. Pappus of capillary bristles.
16. Rays usually more than 50 . . . . 12. *Erigeron*
16. Rays usually fewer than 50.
17. Basal leaves or tufts of leaves usually persistent at flowering time; pappus bristles thickened at tip.
18. Pappus bristles white, in 1 series; inflorescence paniculate; receptacle with small ridges around edge. . . . . . . 1. *Aster*
18. Pappus bristles reddish or yellow, in 3 or 4 series; inflorecence corymbiform; receptacle pitted . . . . 15. *Eurybia*
17. Basal leaves usually absent at flowering time; pappus bristles not thickened at tip . . . . 19. *Symphyotrichum*
15. Pappus of 2 awns and tiny bristles . . . . 14. *Boltonia*
11. Rays white or cream.
19. Pappus of capillary bristles.
20. Pappus in a double series.
21. Inflorescence flat-topped . . . . 3. *Doellingeria*
21. Inflorescence elongated, paniculate. . . . . 4. *Ionactis*
20. Pappus in a single series or in 3 or 4 series.
22. Rays less than 5 mm long . . . . 13. *Conyza*
22. Rays 5 mm long or longer.
23. Rays usually more than 50. . . . . 12. *Erigeron*
23. Rays usually fewer than 50.
24. Basal leaves persistent at flowering time; pappus bristles thickened at tip . . . . 15. *Eurybia*
24. Basal leaves usually absent at flowering time; pappus bristles not thickened at tip.
25. Phyllaries with a translucent midvein; rays 7–9 . . . . 8. *Solidago*
25. Phyllaries without a translucent midvein; rays 9 or more . . . . 19. *Symphyotrichum*
19. Pappus of 2 awns and tiny bristles . . . . 13. *Boltonia*

The sequence of genera follows that in *Flora of North America.*

1. **Aster** L.—Aster

Perennial herbs with short rootstocks, colonial; stems 1 to several, erect to ascending, glabrous or pubescent; basal leaves present at flowering time, large, petiolate; cauline leaves progressively smaller toward top of stem, usually sessile, entire or serrulate, glabrous or pubescent; inflorescence an elongated panicle of rounded or flat-topped clusters; involucre campanulate or hemispheric; phyllaries in 3 or 4 series, unequal, linear to narrowly lanceolate, glabrous or pubescent; receptacle flat, with small ridges around the edge; ray flowers pistillate, blue or white; disc flowers perfect; cypselae obconic, not flat, 4- to 6-nerved, glabrous or pubescent, brown or tan; pappus of numerous white capillary barbellate bristles thickened at the tip, in 1 series.

With the removal of most species in North America usually assigned to *Aster* to other genera, only one non-native species of *Aster* occurs in Illinois. Overall, there are approximately 180 species of *Aster*, all but one of them native to Europe and Asia.

The characteristics of *Aster* are its persistent large, petiolate basal leaves, its flat receptacle with small ridges around the edge, its blue or white rays, and its pappus that is slightly thickened at the tip.

1. **Aster tataricus** L. f. Suppl. Pl. 373. 1782. Fig. 1.

Perennial from thick rhizomes and a thickened caudex, colonial; stems robust, 1 to several, erect to ascending, up to 2 m tall, rough-hairy except near the base, deeply grooved and angular; basal leaves present at flowering time, oblanceolate to elliptic to ovate to ovate-lanceolate, short-acuminate at the apex, more or less rounded at the base, rough-hairy, coarsely serrate, up to 50 cm long, up to 25 cm wide, on winged petioles; cauline leaves progressively smaller, the lower ovate to ovate-lanceolate, acute at the apex, more or less rounded at the base, coarsely serrate, to 40 cm long, to 20 cm wide, the upper narrowly elliptic, subulate-tipped, entire or nearly so, up to 3 cm long, up to 8 mm wide; inflorescence a flat-topped corymb with 8–20 heads on stiffly ascending branches with numerous linear to elliptic bracts; involucre 7–15 mm high, usually as wide; phyllaries in 2–4 series, unequal, linear-lanceolate to oblong, acute to acuminate, glabrous or strigose; ray flowers 15–30, 10–18 mm long, blue or purple; disc flowers up to 50, 4.5–6.0 mm long, yellow; cypselae linear, more or less flat, strigillose, 1.5–2.0 mm long, with 4–6 strong nerves, pale brown; pappus of numerous white or cream capillary bristles 6–8 mm long.

Common Name: Tatarian aster.
Habitat: Disturbed soil.
Range: Native to Siberia; escaped from cultivation and adventive in the United States.
Illinois Distribution: Known from several central counties.

This robust species is sometimes grown as a garden ornamental and is rarely escaped into disturbed soil in Illinois. It is distinguished by its robust stature and its very large basal leaves.

It is the only member of the Illinois flora to remain in the genus *Aster*.

This species flowers during September and October.

1. *Aster tataricus* (Tatarian aster).
a. Habit.
b. Lower leaves.
c. Flowering head.
d. Ray flower.
e. Disc flower.
f. Phyllaries.
g. Cypsela.

2. **Bellis** L.—English Daisy

Perennial herbs; stems sometimes without leaves; leaves basal, usually with a few alternate cauline leaves, simple, entire or toothed; head large, radiate, solitary; involucre hemispheric or campanulate; phyllaries in 1 or 2 series, equal; receptacle epaleate; ray flowers white or pink, pistillate; disc flowers numerous, tubular, yellow, perfect; cypselae flat; pappus absent or merely a ring of minute bristles.

Nine species comprise this genus. Only the following has been found as an adventive in Illinois.

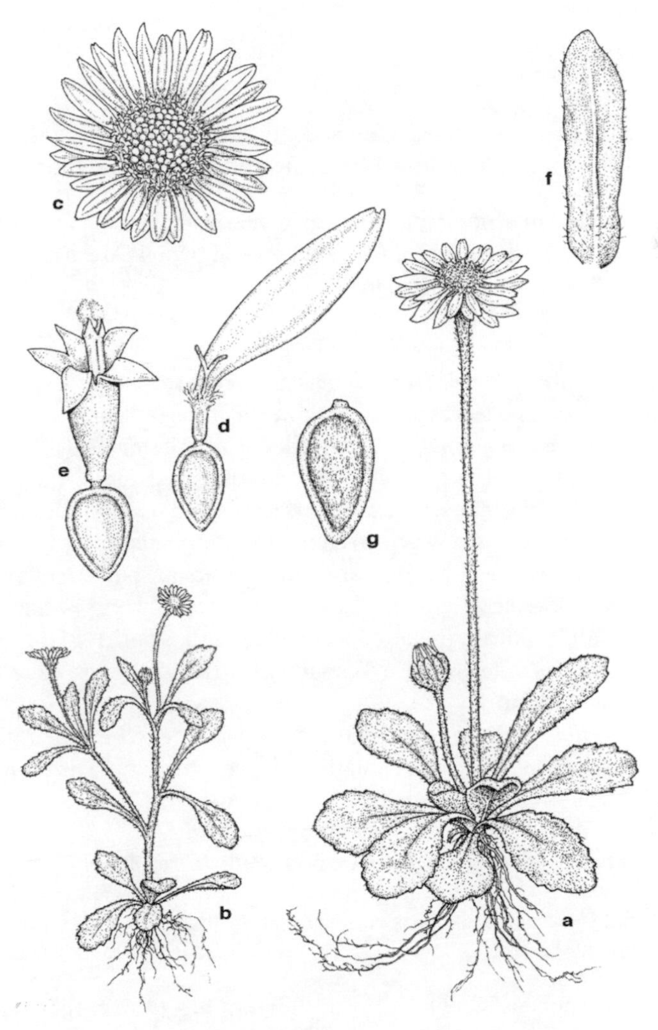

2. *Bellis perennis* (English daisy). a, b. Habit. c. Flowering head. d. Ray flower. e. Disc flower. f. Phyllary. g. Cypsela.

1. **Bellis perennis** L. Sp. Pl. 886. 1753. Fig. 2.

Perennial herb from fibrous roots and short rhizomes; leaves all basal, obovate, obtuse at the apex, tapering to the winged petiolate base, irregularly dentate, pubescent, ciliate, to 4 cm long, to 2.2 cm wide; head usually solitary, to 2.5 cm across, radiate, on a pubescent scape to 15 cm long; involucre to 10 mm high; phyllaries in 1 or 2 series, equal, linear-lanceolate, obtuse at the apex, ciliolate; receptacle epaleate; ray flowers up to 40, white or pink or less commonly purple, linear, to 10 mm long, pistillate; disc flowers numerous, tubular, yellow, perfect, the corolla to 1.5 mm high; cypselae flat, obovate, pubescent, 1–2 mm long; pappus absent.

Common Name: English daisy.
Habitat: Disturbed soil.
Range: Native to Europe; scattered as an adventive in the United States.
Illinois Distribution: Collections from Champaign and McLean counties are known.

This handsome garden ornamental is a rare escape from gardens. Its solitary large flowering head on a leafless scape and the absence of pappus are distinctive.

This species flowers from April to June.

3. **Doellingeria** Nees

Perennial herbs with rhizomes; stems 1 to a few, erect to ascending, striate, glabrous or pubescent; basal leaves absent at flowering time (at least in our species); cauline leaves progressively smaller toward the top of the stem, sessile, glabrous or pubescent, entire or serrulate, often with a submarginal nerve; inflorescence a round- or flat-topped corymb, with numerous heads and usually numerous bracts on the branches; involucres cylindric to campanulate; heads with ray flowers and disc flowers; phyllaries in 2 or 3 series, unequal, lanceolate, pubescent or ciliate; receptacle more or less flat to slightly convex, pitted; ray flowers pistillate, white; disc flowers perfect, yellow; cypselae obovoid, more or less flat, 4- to 6-nerved, pale yellow-brown; pappus of finely barbed capillary bristles in 2 series, usually somewhat broadened at the tip.

Species of *Doellingeria* have often been included in *Aster*. They differ by their round- or flat-topped inflorescences and their pappus in 2 series with the bristles broadened at the tip.

Five species, all in North America, comprise the genus.

1. Phyllaries pubescent; leaves pubescent on both surfaces; ray flowers 4–7 (–12) per head; disc flowers 8–15 (–30) per head . . . . 1. *D. pubens*
1. Phyllaries glabrous; leaves glabrous or nearly so; ray flowers 7–14 per head; disc flowers 30–40 per head . . . . 2. *D. umbellata*

1. **Doellingeria pubens** (Gray) Rydb. Bull. Torrey Club 37:147. 1910. Fig. 3.
Aster umbellatus Mill. var. *pubens* Gray in Gray, Syn. Fl. N. Am. 1:197. 1884.
Doellingeria umbellata (Mill.) Nees var. *pubens* (Gray) Britt. in Britt. & Brown, Ill. Fl. N. U.S. 3:392. 1898.
Aster pubentior Cronq. Bull. Torrey Club 74:147. 1947.

Doellingeria umbellata (Mill.) Nees ssp. *pubens* (Gray) A. Love & D. Love, Taxon 31: 357. 1982.

Perennial from creeping rhizomes; stems 1 to a few, erect to ascending, to 2 m tall, densely pubescent with short curved hairs; basal leaves absent at flowering time, obovate to oblanceolate, acute at the apex, tapering to the sessile base, entire, densely pubescent on both surfaces, entire, the veins forming a submarginal nerve, up to 15 cm long, up to 3.5 cm wide; upper leaves similar but gradually smaller, sessile; inflorescence a flat-topped corymb of 50–150 heads, the heads 1.5–2.5 cm across, the branches with several elliptic leaflike bracts up to 6 cm long; involucre 3.5–5.0 mm high, nearly as wide, more or less turbinate; phyllaries in 3 or 4 series, unequal, narrowly lanceolate to lanceolate, acute at the apex, densely puberulent; ray flowers 4–7 (–12), 5–8 mm long, white; disc flowers 8–15 (–30), yellow; cypselae obovoid, more or less flat, puberulent, 4- or 5-nerved, 1.8–2.5 mm long; pappus a double row of bristles, those of the outer row 0.2–0.5 mm long, cream to tan, the inner 3–4 mm long, finely barbed, cream or tan.

Common Name: Hairy flat-topped aster.
Habitat: Calcareous fens, bogs.
Range: Quebec to Alberta, south to Nebraska, Illinois, and Wisconsin.
Illinois Distribution: Confined to the northern one-sixth of the state.

Although sometimes considered to be a variety of *D. umbellata*, this plant differs by its densely pubescent stems, leaves, and phyllaries, its slightly smaller flowering heads, its fewer ray flowers per head, and its fewer disc flowers per head.

In all specimens of this species I have seen, the ray number per flower is never more than 12, and the disc flowers are never more than 30.

This species flowers from July to October.

2. **Doellingeria umbellata** (Mill.) Nees, Gen. Sp. Aster. 178. 1832. Fig. 4.
Aster umbellatus Mill. Gard. Dict. ed. 8, Aster no. 22. 1768.

Perennial from creeping rhizomes; stems 1 to a few, erect to ascending, to 2.5 m tall, usually glabrous; basal leaves absent at flowering time, obovate to oblanceolate, acute at the apex, tapering to the subsessile base, entire, the lateral veins coming together to form a submarginal nerve, glabrous but often scabrous on the upper surface, glabrous or nearly so on the lower surface, ciliate on the margins, up to 16 cm long, up to 3 cm wide; upper leaves similar but gradually smaller, sessile; inflorescence a flat-topped corymb with 30–300 heads, the heads 2–3 cm across, the branches with several leaflike bracts up to 6 cm long; involucre 2.5–4.5 mm high, nearly as wide; phyllaries in 3 or 4 series, unequal, linear-lanceolate, acute to obtuse at the apex, glabrous, usually ciliate, green throughout; ray flowers 7–14, 5–8 mm long, white; disc flowers 30–40, yellow; cypselae glabrous or sparsely pubescent, 4- to 7-nerved, 1.6–2.2 mm long; pappus a double row of bristles, those of the outer row sordid and thicker at the tip, 0.2–0.8 mm long, those of the inner row 3–4 mm long.

Common Name: Flat-topped aster.
Habitat: Calcareous fens, bogs, low ground.
Range: Newfoundland to Ontario, south to Iowa, Illinois, Alabama, and Georgia.
Illinois Distribution: Occasional in the northern half of Illinois; also Fayette Co.

The more or less glabrous stems, leaves, and phyllaries, the broader flowering heads, and the more numerous ray flowers and disc flowers per head are distinctive from *D. pubens*.

The flowers bloom from July to October.

3. *Doellingeria pubens* (Hairy flat-topped aster).
a. Habit.
b. Leaf.
c. Flowering head.

4. *Doellingeria umbellata* (Flat-topped aster).
a. Upper part of plant.
b. Middle part of plant.
c. Flowering head.
d. Phyllary.
e. Ray flower.
f. Disc flower.
g. Cypsela.

4. **Ionactis** Greene

Perennials (ours) or subshrubs with a woody caudex; stems 1 to several, glabrous or pubescent; leaves all alike, sessile, stiff, linear to narrowly lanceolate, 1-nerved, entire, glabrous or pubescent; inflorescence of solitary heads or 2 or 3 heads in a corymb; heads with ray flowers and disc flowers; involucres turbinate to campanulate; phyllaries in 2–6 series, unequal, keeled, appressed, linear, 1-nerved, glabrous or pubescent; receptacle flat, pitted; ray flowers pistillate, in 1 series, blue or white; disc flowers numerous, perfect, yellow; cypselae narrowly obovoid, flat, usually 3- or 4-nerved, pubescent; pappus in 2 series, the outer of short bristles or scales, the inner of longer bristles, not thickened at the tip.

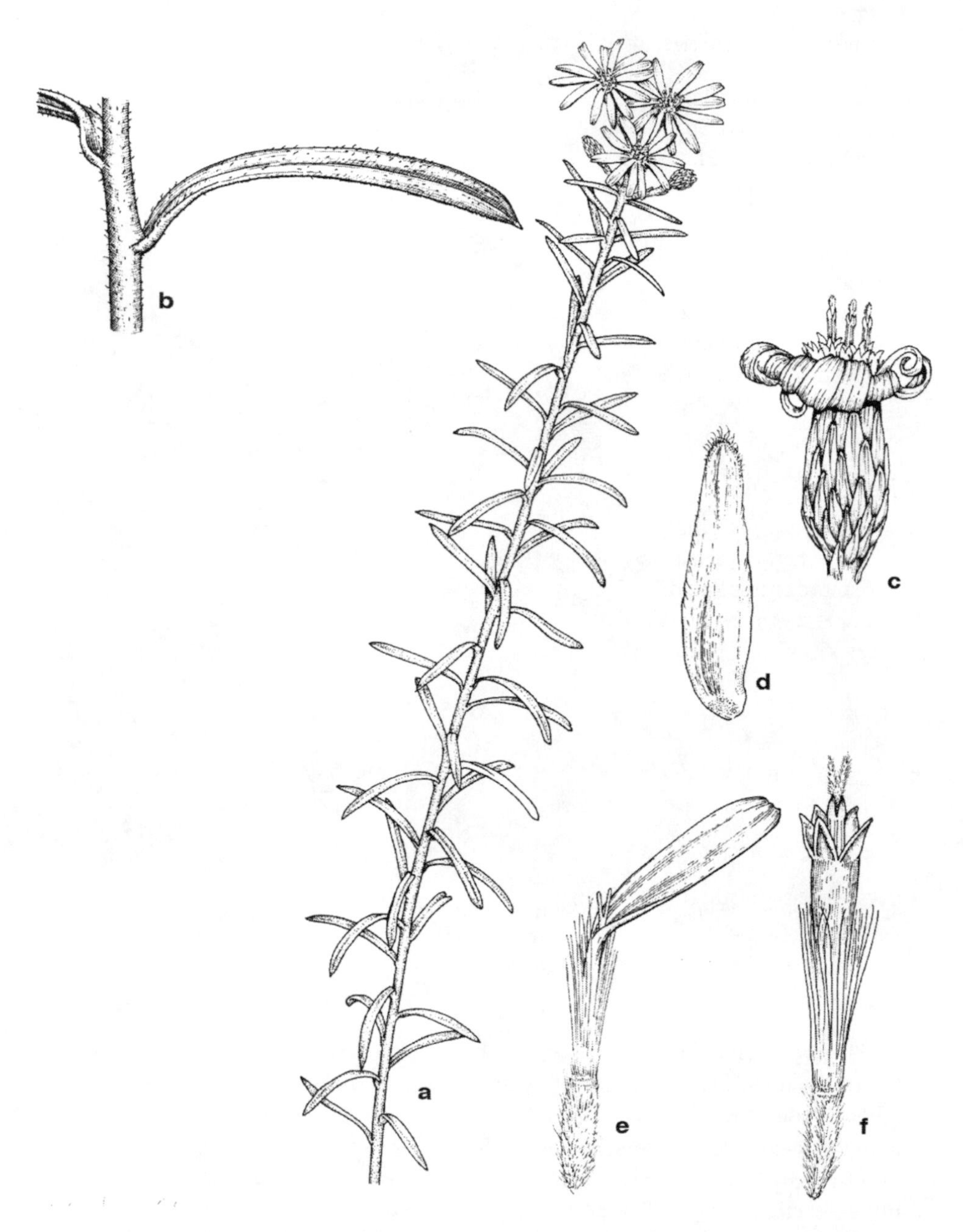

5. *Ionactis linariifolius* (Flax-leaved aster).
a. Habit.
b. Portion of stem with leaves.
c. Flowering head.
d. Phyllary.
e. Ray flower.
f. Disc flower.

The uniform, stiff, narrow leaves, the woody caudices, the few-flowered corymbs, the keeled phyllaries, and the pappus in 2 series of unlike bristles distinguish this genus.

There are five species, all in North America.

1. **Ionactis linariifolius** (L.) Greene, Pittonia 3:245. 1897. Fig. 5.
Aster linariifolius L. Sp. Pl. 2:874. 1753.
Aster rigidus L. Sp. Pl. 2:874. 1753.
Diplopappus linariifolius (L.) Hook. Fl. Bor. Am. 2:21. 1834.

Perennial from a thickened woody caudex; stems 1 to several, erect to ascending, unbranched, to 70 cm tall, glabrous to puberulent to tomentose; basal leaves absent at flowering time; all leaves uniform, stiff, linear to narrowly oblong, acute to subulate at the apex, tapering or rounded at the sessile base, entire, 1-nerved, glabrous and shiny on the upper surface, scabrous and pubescent on the lower surface, often ciliate, up to 4 cm long, up to 4 mm wide; heads solitary or up to 30 in a flat-topped corymb, the branches with numerous linear bracts up to 2 cm long; involucre 6–12 mm high, 5–9 mm wide; phyllaries in 4–7 series, unequal, firm, linear-lanceolate, obtuse to acute, glabrous except for the margin, usually with a greenish tip; ray flowers 6–20, 7–12 mm long, entirely pistillate, lavender to purple, rarely white; disc flowers up to 40, 4–7 mm long, yellow; cypselae obconical, often 4- to 6-angled, more or less flat, densely sericeous, several-nerved, 2.0–3.5 mm long; pappus a double row of bristles, stramineous, the outer ones 0.2–0.8 mm long, the inner series barbed, 4–7 mm long.

Common Name: Flax-leaved aster; stiff aster.
Habitat: Black oak savannas, sandy prairies, sandy barrens.
Range: New Brunswick to Wisconsin, south to Texas and Florida.
Illinois Distribution: Occasional in the northern half of Illinois, south to St. Clair County.

This species is distinguished by its uniform stiff linear leaves, its flat-topped inflorescence, and its pappus in two distinct series. It is often placed in the genus *Aster*.

Ionactis linariifolius flowers from July to October.

5. **Amphiachyris** Nutt.—Broomweed

Glabrous perennial herbs; stems erect, branched; leaves alternate, simple, narrowly linear to lanceolate, entire; heads numerous, small, in corymbs or cymes, radiate; involucres hemispheric to ovoid; phyllaries in several series; receptacle epaleate; ray flowers pistillate, yellow; disc flowers numerous, tubular, perfect but functionally staminate or sometimes sterile, yellow; cypselae terete, pubescent; pappus of ray flowers absent or nearly so, of disc flowers of scales or short bristles.

Two species of the central and western United States and Mexico comprise the genus. Only the following has been found in Illinois.

1. **Amphiachyris dracunculoides** (DC.) Nutt. Trans. Am. Phil. Soc. n.s. 7:313. 1840. Fig. 6.

Brachyris dracunculoides DC. Mem. Soc. Phys. Gen. 7:265. 1836.

Gutierrezia dracunculoides (DC.) S. F. Blake, Contr. U.S. Natl. Herb. 22:592. 1924.

Annual herb from fibrous roots; stems erect to ascending, much branched, to 50 cm tall; leaves alternate, simple, linear to narrowly lanceolate, the uppermost nearly filiform, acute at the apex, tapering to the sessile base, entire, glabrous, to 5.5 cm long, to 6 mm wide; heads numerous, crowded in corymbs; involucres hemispheric; phyllaries in a few series, the outer ones shorter; receptacle epaleate; ray flowers 5–10, 2–3 mm long, yellow, pistillate; disc flowers several, yellow, perfect but functionally staminate or even sterile; cypselae terete, 1.2–2.2 mm long, several-ribbed, short-pubescent; pappus of ray flowers absent or nearly so, of disc flowers of scales or short bristles.

6. *Amphiachyris dracunculoides* (Prairie broomweed). a, b. Habit. c. Flowering head, side view. d. Flowering head, face view. e. Phyllary. f. Ray flower. g. Disc flower. h. Cypsela.

Common Name: Prairie broomweed.
Habitat: Disturbed soil (in Illinois).
Range: Iowa to Nebraska, south to New Mexico and Oklahoma, and scattered in several states.
Illinois Distribution: Adventive in Jefferson and Tazewell counties.

This is a common species of the plains and prairies west of Illinois. Its two stations in Illinois have persisted for several years.

Sometimes this species is placed in the genus *Gutierrezia.*

This species flowers from July to October.

6. **Gutierrezia** Lagasca—Snakeweed

Glabrous herbs or shrubs; stems erect, usually branched above; leaves alternate, simple, linear to narrowly lanceolate, entire; heads numerous, small in corymbs or panicles, radiate; involucre hemispheric to campanulate; phyllaries in several series; receptacle epaleate; ray flowers few, yellow, pistillate; disc flowers several, yellow, tubular, perfect of sometimes staminate only; cypselae terete, sparsely pubescent; pappus of ray flowers shorter or equaling pappus of disc flowers, composed of scales or awns.

Twenty-eight species comprise this genus in the central and western United States, Mexico, and South America.

Only the following occurs in Illinois.

1. **Gutierrezia texana** (DC.) Torr. & Gray, Fl. N. Am. 2:194. 1842. Fig. 7.
Hemiachyris texana DC. in A.P. DC. & A.L.P.P. DC. Prodr.5:314. 1836.
Xanthocephalum texanum (DC.) Shinners, Field & Lab. 18:28. 1950.

Annual herbs from fibrous roots; stems erect, to nearly 1 m tall, branched at least above, glabrous; leaves alternate, simple, linear, acute at the apex, entire, glabrous, 1- or 3-nerved, up to 4 mm wide; heads several in corymbs, to 5 mm across; involucre campanulate, 2–3 mm across; phyllaries in several series; receptacle epaleate; ray flowers up to about 20, yellow, fertile, 3–6 mm long; disc flowers up to 15, yellow, tubular, fertile or only staminate; cypselae terete, 1.3–1.8 mm long, 5-angled, strigose; pappus of several series, those of the ray flowers shorter than those of the disc flowers, or pappus absent.

Common Name: Texas snakeweed.
Habitat: Disturbed soil (in Illinois).
Range: Missouri to New Mexico, east to Louisiana.
Illinois Distribution: Adventive in Madison and St. Clair counties.

This species, common in the southwestern United States, has been found once in disturbed soil in Madison County. It differs from *Amphiachyris dracunculoides*, which it resembles, by having fertile disc flowers.

The flowers appear from July to October.

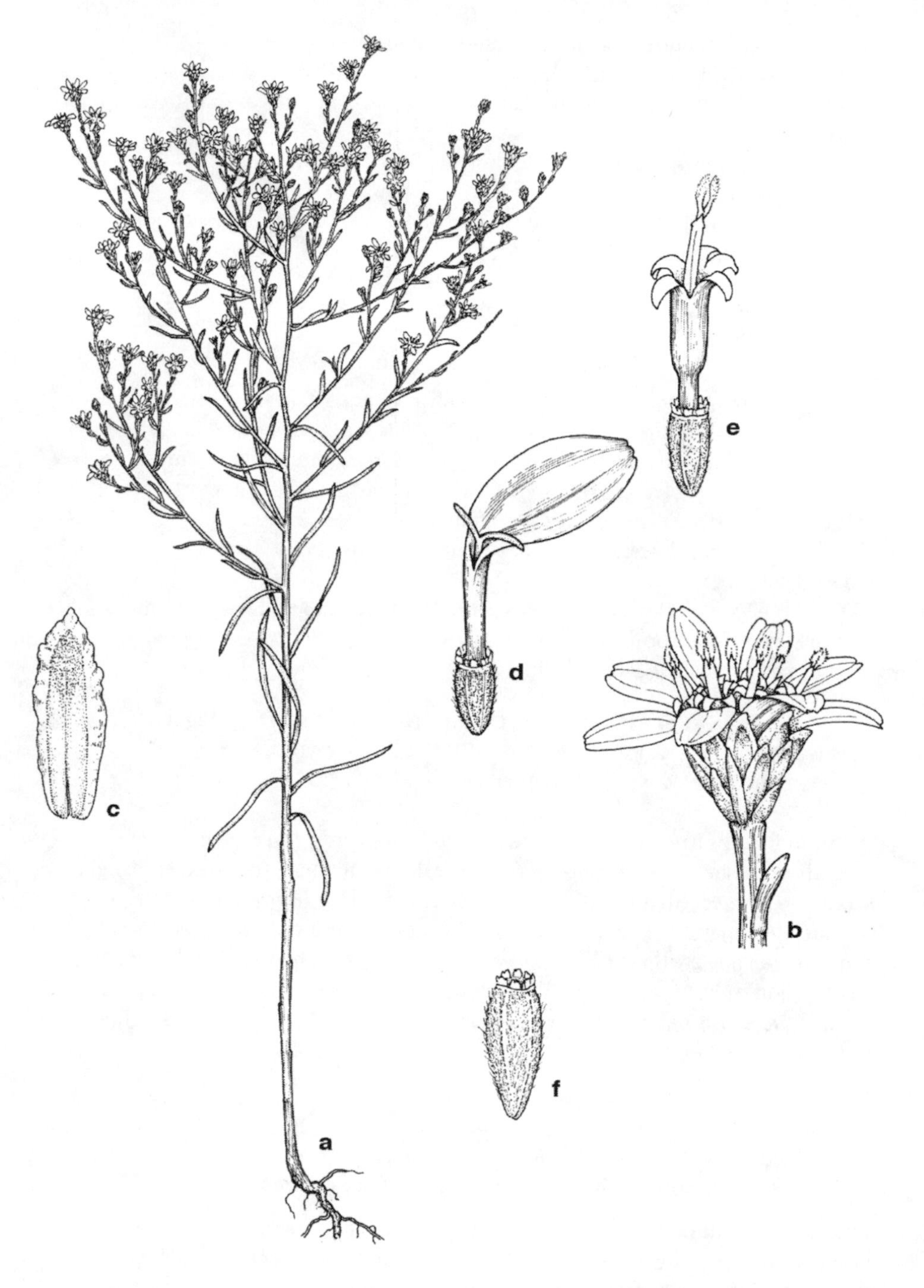

7. *Gutierrezia texana* (Texas snakeweed).

a. Habit.
b. Flowering head.
c. Phyllary.
d. Ray flower.
e. Disc flower.
f. Cypsela.

7. **Euthamia** (Nutt.) Cass.—Grass-leaved Goldenrod

Perennial herbs with rhizomes and usually a woody caudex; stems usually solitary, branched, erect to ascending, striate, glabrous or pubescent; basal leaves absent; cauline leaves smaller near top of stem, all leaves linear to linear-lanceolate to lanceolate, acute at the apex, tapering to the sessile base, entire but often stiff-ciliate, glabrous or pubescent, with 1, 3, or 5 parallel veins, glandular-dotted although sometimes obscurely so; inflorescence a flat-topped corymb, the heads 1 to several in glomerules, radiate; involucre obconic to turbinate to campanulate; phyllaries in 4 or 5 series, unequal, appressed, the outer often ovate, the inner often linear to linear-lanceolate, acute to obtuse, glabrous or sometimes resinous, green-tipped; receptacle flat, with toothed ridges and sometimes pubescent; ray flowers pistillate, 1–3 mm long, usually shorter than the disc flowers, yellow; disc flowers perfect, yellow; cypselae oblongoid, usually not flat, obscurely ribbed, pubescent, stramineous to greenish brown; pappus of numerous white capillary bristles.

Euthamia consists of five species, all in North America. Some botanists recognize additional species. Four species have been found in Illinois.

1. Leaves up to 3 (–8) mm wide, some of them 1-nerved, with numerous obvious glandular dots.
   2. Involucre 3.0–4.7 mm high; leaves never more than 3 mm wide; disc flowers 3–22 . . . . 1. *E. caroliniana*
   2. Involucre 4.5–6.0 mm high; leaves often more than 3 mm wide, up to 8 mm wide; disc flowers up to 9. . . . . 2. *E. gymnospermoides*
1. Leaves 3–12 mm wide, all of them 3- or 5-nerved, with a few obscure glandular dots, or eglandular.
   3. Stems and leaves often densely hirtellous; inner phyllaries oblong; ray flowers up to 35 in number; disc flowers up to 13 in number, the corolla 2.5–3.3 mm long . . . . 3. *E. graminifolia*
   3. Stems and leaves glabrous; inner phyllaries linear; ray flowers up to 14 in number; disc flowers up to 6 in number, the corolla 3.5–4.5 mm long . . . . . . 4. *E. leptocephala*

1. **Euthamia caroliniana** (L.) Greene ex Porter & Britt. Mem. Torrey Club 5:321. 1894. Fig. 8.

Erigeron carolinianus L. Sp. Pl. 2:863. 1753.
Solidago lanceolata L. var. *minor* Michx. Fl. Bor. Am. 2:116. 1803.
Solidago tenuifolia Pursh, Fl. Am. Sept. 2:540. 1813.
Euthamia tenuifolia (Pursh) Nutt. Gen. N. Am. Pl. 2:162. 1818.
Euthamia remota Greene, Pittonia 5:78. 1902.
Solidago remota (Greene) Friesn. Butler Univ. Bot. Stud. 3:62. 1933.
Solidago graminifolia (L.) Salisb. var. *remota* (Greene) S. K. Harris, Rhodora 45:413. 1943.

Perennial herbs with creeping rhizomes; stems 1 to several, ascending to erect, branched, glabrous or nearly so, to 1 m tall; all leaves cauline, linear, mostly 1-nerved, acute to acuminate at the apex, tapering to the sessile base, to 7 mm long, to 3 mm wide, glabrous or nearly so, strongly and profusely glandular-dotted; inflorescence round- or flat-topped, with numerous heads 3.5–5.5 mm across; involucre

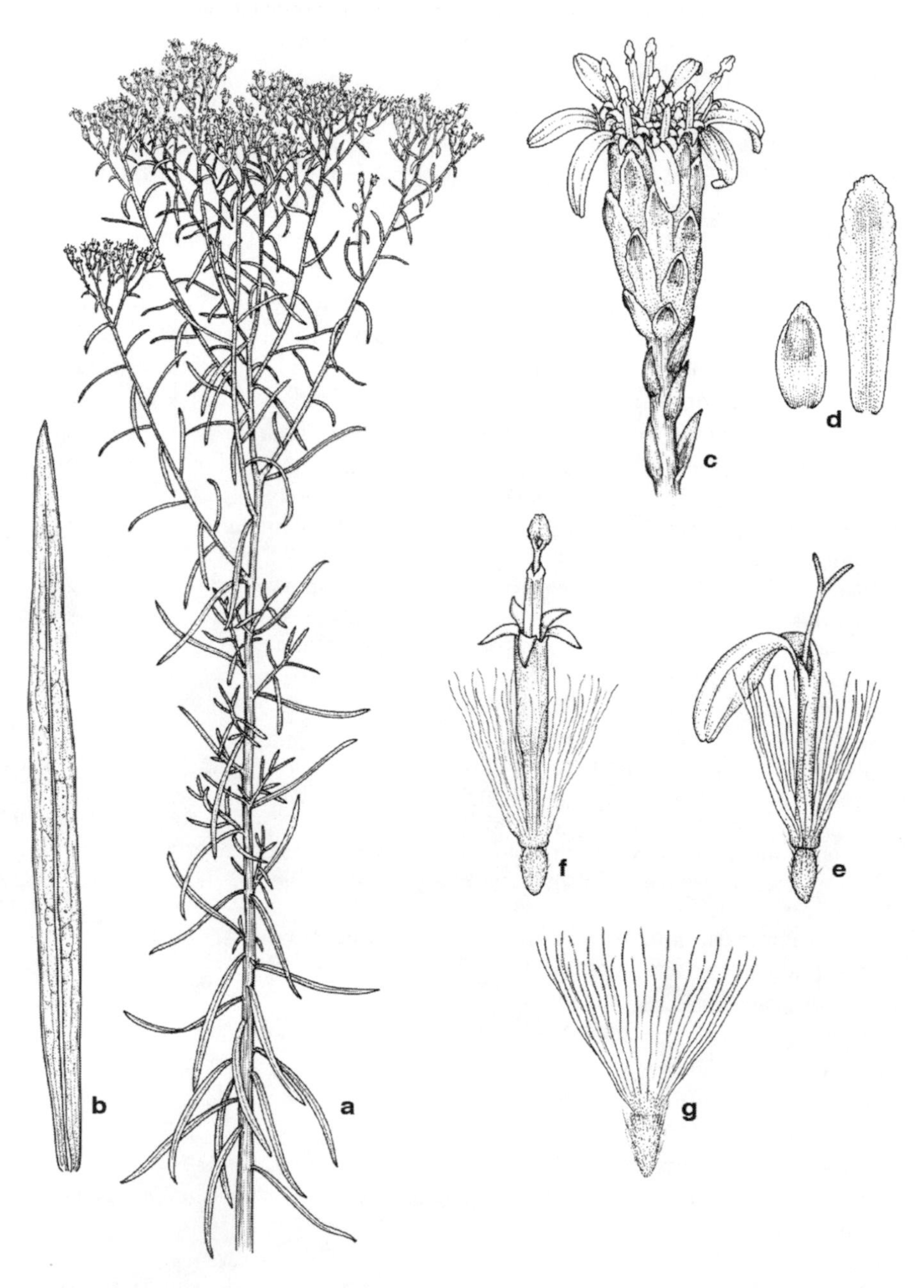

8. *Euthamia caroliniana* (Carolina grass-leaved goldenrod).

a. Habit.
b. Leaf.
c. Flowering head.
d. Phyllaries.
e. Ray flower.
f. Disc flower.
g. Cypsela.

campanulate to turbinate, 3.0–4.7 mm high, usually viscid; phyllaries usually in 4 series, green-tipped, the outer ovate, acute, the inner narrowly oblong, usually obtuse; receptacle flat, with toothed ridges; ray flowers 7–25, yellow, 1–3 mm long; disc flowers 3–22, yellow, the corolla 3.0–4.7 mm long; cypselae oblongoid to ellipsoid, 1–2 mm long, few-nerved, more or less pubescent; pappus of numerous white capillary bristles 4–5 mm long in 1 series.

Common Name: Carolina grass-leaved goldenrod.
Habitat: Sandy soil in prairies.
Range: Maine to Wisconsin, south to Illinois, Louisiana, and Florida.
Illinois Distribution: Known only from Cook, Lake, and Kankakee counties.

The strongly and prominently glandular-dotted leaves are similar to those of *E. gymnospermoides*, but *E. caroliniana* has narrower leaves and more numerous disc flowers.

If our plant is different from the plants in the southeastern part of the United States, it should be called *E. remota*.

Euthamia caroliniana flowers during August and September.

2. **Euthamia gymnospermoides** Greene, Pittonia 5:75. 1902. Fig. 9.
Solidago gymnospermoides (Greene) Fern. Rhodora 10:93. 1908.

Resinous perennial herbs from creeping rhizomes; stems 1 to several, ascending to erect, to 1.5 m tall, usually branched, glabrous; all leaves cauline, ascending, linear to linear-lanceolate, acuminate at the apex, tapering to the sessile base, 1-, 3-, or 5-nerved, to 8 (–10) cm long, 2–5 (–8) mm wide, glabrous except for the midvein below, strongly and profusely glandular-dotted; inflorescence round-topped, with numerous heads 4–6 mm across; involucre obconic, 4.5–6.0 mm high, viscid; phyllaries usually in 5 series, green-tipped, the outer ovate, acute, the inner narrowly oblong, usually obtuse, resinous; receptacle flat, with toothed ridges; ray flowers 9–15, yellow, 1–3 mm long; disc flowers 3–9, yellow, the corolla 3–5 mm long; cypselae oblongoid to ellipsoid, 1–2 mm long, few-nerved, more or less strigose; pappus of numerous white capillary bristles 4–5 mm long in 1 series.

Common Name: Viscid grass-leaved goldenrod.
Habitat: Prairies, sandy soil.
Range: Ontario to Minnesota to South Dakota and Colorado, south to Texas and Florida.
Illinois Distribution: Occasional in northern and western Illinois.

Because of its prominently glandular-dotted leaves, this species is similar to *E. caroliniana*, differing by its longer disc flowers and its longer involucres.

Euthamia gymnospermoides flowers from August to October.

9. *Euthamia gymnospermoides* (Viscid grass-leaved goldenrod).

a, b. Habit.
c. Flowering head.
d. Phyllary.

e. Ray flower.
f. Disc flower.
g. Cypsela.

3. **Euthamia graminifolia** (L.) Nutt. Trans. Am. Phil. Soc. n.s. 7:325. 1840.
Chrysocoma graminifolia L. Sp. Pl. 2:841. 1753.
Solidago graminifolia (L.) Salisb. Prodr. Stirp. Chap. Allerton 199. 1796.
Euthamia media Greene, Pittonia 5:74–75. 1902.

Perennial herbs from creeping rhizomes; stems ascending to erect, to 1.5 m tall, branched, glabrous or densely hirtellous; all leaves cauline, linear to lanceolate, acute to acuminate at the apex, tapering to the sessile base, 3- or 5-nerved, glabrous or densely hirtellous, to 12 cm long, to 12 mm wide, obscurely glandular-dotted; inflorescence flat-topped, with numerous heads 4–6 mm in diameter; involucre campanulate, 3.0–5.5 mm high; phyllaries in 4 or 5 series, green-tipped, the outer ovate, acute, the inner oblong, acute to obtuse; receptacle flat, with toothed ridges; ray flowers up to 35 per head, yellow, 1–3 mm long; disc flowers up to 13 per head, yellow, the corolla 2.5–3.3 mm high; cypselae oblongoid to ellipsoid, 1–2 mm long, few-nerved, more or less pubescent; pappus of numerous white capillary bristles 4–5 mm long in 1 series.
Two varieties occur in Illinois:
a. Leaves and stems glabrous or nearly so . . . . 3a. *E. graminifolia* var. *graminifolia*
a. Leaves and stems densely hirtellous . . . . 3b. *E. graminifolia* var. *nuttallii*

3a. **Euthamia graminifolia** (L.) Nutt. var. **graminifolia**
Leaves and stems glabrous or nearly so.

Common Name: Grass-leaved goldenrod.
Habitat: Moist ground, fields, sandy shores, calcareous areas.
Range: All across Canada, south to Oregon, Colorado, Oklahoma, Mississippi, and South Carolina.
Illinois Distribution: Common throughout the state.

This variety appears to be the more common one in Illinois. Plants with narrow leaves and acute inner phyllaries have been called *E. media* Greene.
Euthamia graminifolia var. *graminifolia* flowers from August to October.

3b. **Euthamia graminifolia** (L.) Nutt. var. **nuttallii** (Greene) Sieren, Rhodora 83:564. 1981. Fig. 10.
Euthamia nuttallii Greene, Pittonia 5:73. 1902.
Euthamia hirtella Greene, Leafl. Bot. Obs. 1:180. 1906.
Solidago graminifolia (L.) Nutt. var. *nuttallii* (Greene) Fern. Rhodora 10:92. 1908.
Solidago nuttallii (Greene) Bush, Am. Midl. Nat. 5:168. 1918.
Solidago hirtella (Greene) Bush ex Friesn. Butler Univ. Bot. Stud. 3:59. 1933.
Solidago perglabra Friesn. Butler Univ. Bot. Stud. 3:61. 1933.

Leaves and stems hirtellous.

Common Name: Grass-leaved goldenrod.
Habitat: Moist ground.
Range: Newfoundland to Minnesota, south to Missouri and North Carolina.
Illinois Distribution: Scattered throughout Illinois.

Although sometimes considered to be a distinct species, this plant appears to be treated best as a variety of *E. graminifolia*. Its main differences are the hirtellous stems and leaves.

Friesner's *Solidago perglabra*, the type of which is from Watseka in Iroquois County, is a synonym for this species.

This variety flowers from August to October.

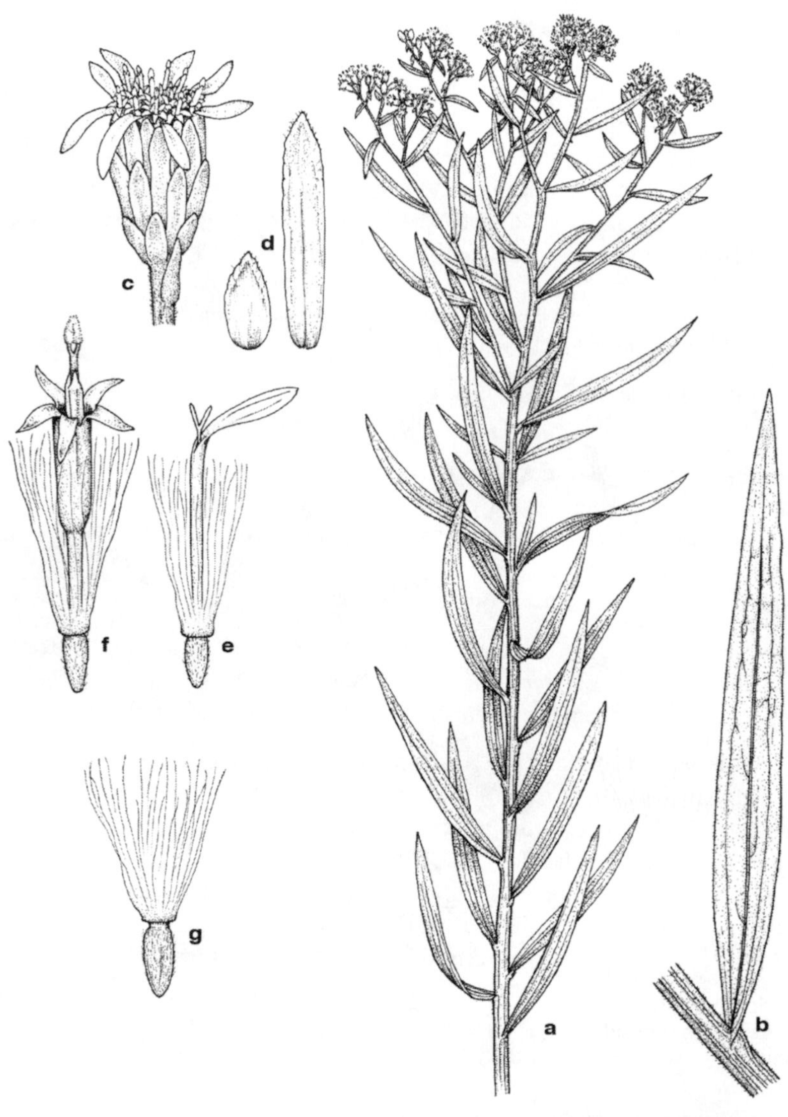

10. *Euthamia graminifolia* var. *nuttallii* (Grass-leaved goldenrod).

a. Habit.
b. Leaf and node.
c. Flowering head.
d. Phyllaries.
e. Ray flower.
f. Disc flower.
g. Cypsela.

4. **Euthamia leptocephala** (Torr. & Gray) Greene ex Porter & Britt. Mem. Torrey Club 5:321. 1894. Fig. 11.
Solidago leptocephala Torr. & Gray, Fl. N. Am. 2:226. 1842.

Nonresinous perennial herbs from creeping rhizomes; stems 1 to several, ascending to erect, to 1 m tall, branched above, glabrous; all leaves cauline, spreading

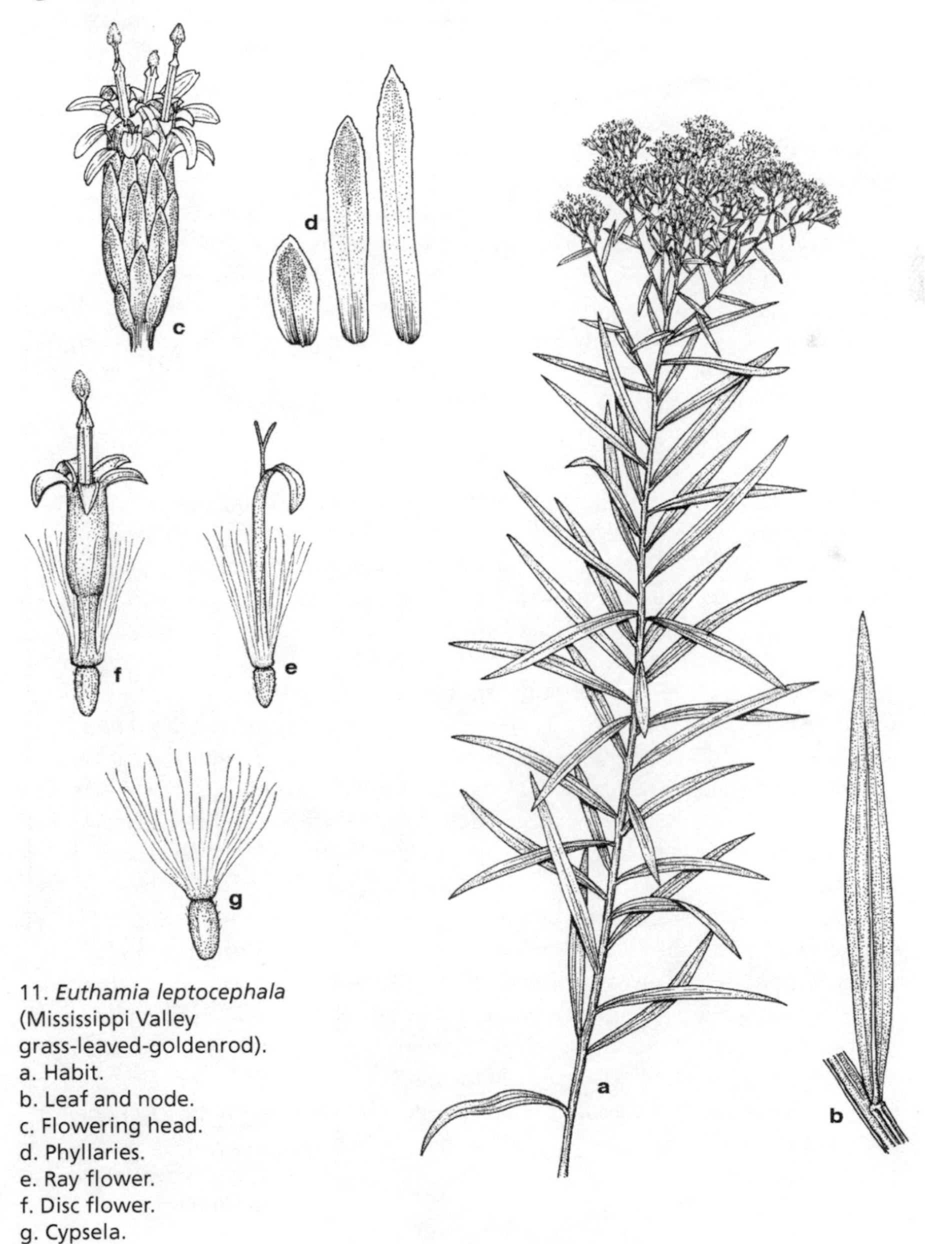

11. *Euthamia leptocephala* (Mississippi Valley grass-leaved-goldenrod).
a. Habit.
b. Leaf and node.
c. Flowering head.
d. Phyllaries.
e. Ray flower.
f. Disc flower.
g. Cypsela.

to ascending, linear-lanceolate to lanceolate, acute at the apex, tapering to sessile base, with 3 or 5 obscure nerves, to 8 cm long, 3–6 (–8) mm wide, glabrous, obscurely glandular-dotted; inflorescence round-topped, with fewer than 20 heads 4–6 mm across; involucre obconic to turbinate, 4.5–6.5 mm high; phyllaries in 4 series, green-tipped, the outer narrowly ovate, acute, the inner linear, obtuse; receptacle flat, with toothed ridges; ray flowers 7–12 (–14), yellow, 1–3 mm long; disc flowers 3–5 (–6), yellow, the corolla 3.5–4.5 mm long; cypselae oblongoid to ellipsoid, 1–2 mm long, few-nerved, strigose; pappus of 20–30 white capillary bristles 4–5 mm long in 1 series.

Common Name: Mississippi Valley grass-leaved goldenrod.
Habitat: Sandy woods.
Range: North Carolina to Missouri, south to Texas and Florida.
Illinois Distribution: Confined to the southern one-sixth of the state.

This species is similar in appearance to *E. gymnospermoides* but is scarcely or not at all resinous. Most of the few specimens of this plant are from the southwestern corner of the state.

Euthamia leptocephala flowers from August to October.

8. **Solidago** L.—Goldenrod

Perennial herbs from rhizomes and often a woody caudex; stems 1 to several, erect to ascending to spreading, glabrous or pubescent; basal leaves usually absent at flowering time, the cauline leaves alternate and progressively smaller toward the top of the stem, tapering or rounded at the base, rarely clasping, entire to serrate; inflorescence paniculate or in thyrses, the branches ascending to spreading, with several heads or with the head solitary, with small, often leaflike bracts, or the inflorescence from the axils of some of the cauline leaves; heads radiate; involucre campanulate to cylindric; phyllaries in 3–6 series, unequal, appressed or less commonly squarrose, usually ovate with 1 vein, glabrous or pubescent; receptacle flat, with toothed ridges, paleate or epaleate; ray flowers pistillate, yellow or rarely cream; disc flowers perfect, yellow; cypselae ovoid or obovoid, not flat, several-nerved, glabrous or pubescent, stramineous or brown; pappus of uniform capillary bristles, white or nearly so.

About 100 species are in this genus, most of them native to North America.

The sequence of species follows that proposed in *Flora of North America* (2006). *Flora of North America* divides *Solidago* into two Sections—*Solidago* and *Ptarmicoidei*.

In this work, I am placing the species in Section *Ptarmicoidei* in the genus *Oligoneuron*. In addition, some species that used to be placed in *Solidago* I am accepting as belonging to the genus *Euthamia*. I am using the breakdown into subsections and series of Section *Solidago* used in *Flora of North America*. The sequence follows:

Subsection **Multiradiatae**. There are no species in Illinois.

Subsection **Humiles**. There are no species in Illinois, although the following taxon should be looked for in the state: *S. simplex* var. *gilmanii*.

Subsection **Thyrsiflorae**.

1. *Solidago petiolaris*
2. *Solidago angusta*
3. *Solidago buckleyi*

Subsection **Squarrosae**. In addition to the species listed below, *S. erecta* may be found in Illinois eventually, since it is known from Indiana and Kentucky where it occurs in dry woods.

4. *Solidago bicolor*
5. *Solidago hispida*
6. *Solidago speciosa*
7. *Solidago rigidiuscula*
8. *Solidago jejunifolia*
9. *Solidago sciaphila*

Subsection **Glomeruliflorae**.

10. *Solidago caesia*
11. *Solidago flexicaulis*

Subsection **Argutae**, Series **Argutae**.

12. *Solidago arguta*
13. *Solidago boottii*
14. *Solidago patula*

Subsection **Argutae**, Series **Brachychaetae**.

15. *Solidago sphacelata*

Subsection **Maritimae**.

16. *Solidago sempervirens*
17. *Solidago uliginosa*
18. *Solidago purshii*

Subsection **Junceae**. In addition to the species listed below, *S. gattingeri* should be expected on limestone glades along the Mississippi River from Calhoun County south to Randolph County. It is known from glades across the Mississippi River in Missouri.

19. *Solidago juncea*
20. *Solidago glaberrima*

Subsection **Venosae**, Series **Venosae**.

21. *Solidago ulmifolia*
22. *Solidago rugosa*
23. *Solidago aspera*

Subsection **Venosae**, Series **Odoreae**. No species in this series is known from Illinois yet, but *S. odora* is expected in dry woods in several of the southern counties. It is known from adjacent counties in Missouri and Kentucky.

Subsection **Venosae**, Series **Drummondiani**.

24. *Solidago drummondii*

Subsection **Triplinerviae**. In addition to the species listed below, *S. rupestris* may eventually be found along riverbanks in southeastern Illinois. It is known from Indiana and Kentucky.

25. *Solidago canadensis*
26. *Solidago altissima*
27. *Solidago gigantea*

Subsection **Nemorales**. In addition to the species listed below, *S. mollis* may eventually be found in prairies in northwestern Illinois. It is known from Wisconsin and Iowa.

28. *Solidago nemoralis*
29. *Solidago decemflora*
30. *Solidago radula*

Stems glabrous below the inflorescence: *arguta, caesia, flexicaulis, gigantea, glaberrima, jejunifolia, juncea, patula, purshii, rigidiuscula, sciaphila, sempervirens, speciosa, uliginosa, ulmifolia.*

Stems pubescent below the inflorescence: *altissima, angusta, arguta, aspera, bicolor, boottii, buckleyi, canadensis, decemflora, drummondii, hispida, nemoralis, patula* (rarely), *petiolaris, radula, rugosa.*

Basal leaves present at flowering time: *arguta, bicolor, boottii, decemflora, hispida, jejunifolia, nemoralis, patula, purshii, sciaphila, sempervirens, speciosa, sphacelata, uliginosa.*

Basal leaves absent at flowering time: *altissima, angusta, aspera, buckleyi, caesia, canadensis, drummondii, flexicaulis, gigantea, glaberrima, juncea, petiolaris, radula, rigidiuscula, rugosa, ulmifolia.*

Cypselae up to 2 mm long: *altissima, arguta, aspera, bicolor, buckleyi, caesia, canadensis, decemflora, drummondii, flexicaulis, gigantea, glaberrima, hispida, jejunifolia, juncea, patula, purshii, radula, rigidiuscula, rugosa, sempervirens, speciosa, sphacelata, uliginosa.*

Cypselae more than 2 mm long: *angusta, arguta, bicolor, boottii, buckleyi, caesia, decemflora, drummondii, nemoralis, petiolaris, radula, sciaphila.*

Cypselae glabrous: *angusta, arguta, bicolor, buckleyi, gigantea, glaberrima, hispida* (except when immature), *jejunifolia, juncea, patula* (rarely), *petiolaris, purshii, rigidiuscula, speciosa, uliginosa.*

Cypselae pubescent: *altissima, arguta, aspera, boottii, caesia, canadensis, decemflora, drummondii, flexicaulis, gigantea, glaberrima, juncea, nemoralis, patens, radula, rugosa, sciaphila, sempervirens, ulmifolia.*

Flowering heads secund: *altissima, arguta, aspera, boottii, canadensis, decemflora, drummondii, gigantea, glaberrima, juncea, nemoralis, patula, radula, rugosa, sempervirens, sphacelata, uliginosa, ulmifolia.*

Flowering heads not secund: *angusta, bicolor, buckleyi, caesia, flexicaulis, hispida, jejunifolia, purshii, rigidiuscula, speciosa, uliginosa.*

Leaves conspicuously 3-veined above the base: *altissima, canadensis, drummondii, gigantea, glaberrima, radula, sempervirens.*

Leaves 1-veined: *angusta, arguta, aspera, bicolor, boottii, buckleyi, caesia, decemflora, flexicaulis, hispida, jejunifolia, juncea, nemoralis, patula, petiolaris, purshii, rigidiuscula, rugosa, sciaphila, sphacelata, speciosa, uliginosa, ulmifolia.*

Flowering heads in axillary clusters or forming a thyrse: *angusta, bicolor, buckleyi, caesia, flexicaulis, hispida, petiolaris, purshii, rigidiuscula, sciaphila, speciosa, uliginosa.*

Flowering heads in panicles: *altissima, arguta, aspera, boottii, canadensis, decemflora, drummondii, gigantea, glaberrima, jejuniflora, juncea, nemoralis, patula, radula, rugosa, sempervirens, sphacelata, uliginosa, ulmifolia.*

Leaves or phyllaries glutinous: *angusta* (leaves), *rigidiuscula* (phyllaries).

Stems glaucous: *caesia, gigantea, patula.*

Leaves cordate: *sphacelata.*

Leaves sweet-scented: *jejunifolia.*

Stems angled: *flexicaulis, patula.*

Pappus less than 1 mm long: *sphacelata.*

There are several complexes of species of *Solidago* that appear to form more or less a continuum in the United States. These include *S. altissima–S. canadensis, S. nemoralis–S. decemflora, S. uliginosa–S. purshii, S. petiolaris–S. angusta, S. arguta–S. boottii, S. rugosa–S. aspera, and S. speciosa–S. rigidiuscula–S. jejunifolia.* After studying the morphological differences of the members of these complexes in the field for more than 60 years, I have come to the conclusion that each member of the complexes mentioned above should deserve species status. While there are some specimens that prove difficult to determine, this phenomenon is found throughout many species complexes in the United States in most families.

Key to the Species of *Solidago* in Illinois

1. Flower heads in axillary clusters and sometimes forming a thyrse above; flower heads not secund.
  2. Stems glabrous, at least below the inflorescence.
    3. Flowering heads only in axillary clusters; leaves subtending all but the uppermost flower clusters longer than the clusters.
      4. Stems strongly glaucous, terete, often arching, not zigzag; leaves lanceolate to elliptic, to 3 cm wide . . . . 10. *S. caesia*
      4. Stems not glaucous, often with ridges, usually erect, often zigzag; leaves broadly ovate to broadly lanceolate, to 10 cm wide . . . . 11. *S. flexicaulis*
    3. Uppermost flowering heads forming a thyrse; the lowest cluster of flower heads subtended by leaves shorter than the clusters.

5. Basal leaves withered at flowering time; cauline leaves less than 2.5 cm wide . . . . 7. *S. rigidiuscula*
5. Basal leaves persistent at flowering time; cauline leaves usually 2.5 cm wide or wider.
6. Cypselae pubescent, 2.5–3.0 mm long. . . . . 9. *S. sciaphila*
6. Cypselae glabrous, up to 2.5 (–2.6) mm long.
7. Lowest leaves at least seven times longer than wide; disc flowers 9–15; petioles with a sheathing base . . . . 18. *S. purshii*
7. Lowest leaves less than seven times longer than wide; disc flowers 4–10; petioles without a sheathing base . . . . 6. *S. speciosa*
2. Stems pubescent below the inflorescence, although sometimes glabrous near the base.
8. Basal leaves persistent at flowering time, larger than the middle and upper leaves.
9. Rays white . . . . 4. *S. bicolor*
9. Rays yellow. . . . . 5. *S. hispida*
8. Basal leaves absent at flowering time; before anthesis, basal leaves smaller than the middle and upper cauline leaves.
10. Leaves thin, serrate; cypselae 2.0–2.7 mm long; phyllaries appressed . . . . 3. *S. buckleyi*
10. Leaves thick and firm, sparsely serrate to entire; cypselae 3–4 mm long; outer phyllaries squarrose.
11. Leaves and phyllaries glutinous; leaves less than 3 cm wide; stems without spreading hairs . . . . 2. *S. angusta*
11. Leaves and phyllaries not glutinous; some or all the leaves more than 3 cm wide; stems with spreading hairs . . . . 1. *S. petiolaris*
1. Flower heads in terminal pyramidal panicles, usually with spreading branches; flower heads usually secund.
12. Stems pubescent, at least below the inflorescence.
13. Basal and lower cauline leaves cordate; pappus up to 1 mm long, shorter than the cypsela . . . . 15. *S. sphacelata*
13. None of the leaves cordate; pappus more than 1 mm long, as long as or longer than the cypsela.
14. Leaves prominently 3-veined above the base, the basal absent at flowering time.
15. Leaves ovate to elliptic; cypselae 1.5–2.5 mm long.
16. Leaves rigid, elliptic; all except the uppermost leaves petiolate, scabrous . . . . 30. *S. radula*
16. Leaves not rigid, broadly ovate to broadly elliptic; middle and upper leaves sessile, softly pubescent . . . . 24. *S. drummondii*
15. Leaves lanceolate to oblanceolate; cypselae 0.8–1.5 mm long.
17. Involucre 3.0–4.5 mm high; ray flowers 10–16, 3–4 mm long; disc flowers 3–7, 3.0–3.5 mm long. . . . . 26. *S. altissima*
17. Involucre 2–3 mm high; ray flowers 6–12, 2–3 mm long; disc flowers 2–5, 2.3–2.7 mm high . . . . 25. *S. canadensis*
14. Leaves 1-veined, the basal usually present at flowering time (except *S. aspera* and *S. rugosa*).
18. Basal leaves absent at flowering time.
19. Stems densely hispid; leaves scabrous, thick and firm, strongly rugose, the teeth more or less blunt. . . . . 22. *S. aspera*

19. Stems spreading-villous; leaves villous, thin, scarcely rugose, the teeth sharply pointed . . . . 23. *S. rugosa*

18. Basal leaves present at flowering time.

20. Lower leaves broadly ovate to obovate to broadly elliptic, not gray-puberulent.

21. Stems strongly angled and sometimes slightly winged; leaves harshly scabrous on the upper surface . . . . 14. *S. patula*

21. Stems terete or nearly so, unwinged; leaves glabrous or slightly scabrous on the upper surface.

22. Leaves rather thin, glabrous or slightly scabrous above, sparsely pubescent or glabrous below; cypselae short-hispid . . . . 12. *S. arguta*

22. Leaves thick, glabrous above, densely pubescent below; cypselae glabrous or sparsely pubescent . . . . 13. *S. boottii*

20. Lower leaves oblanceolate, gray-puberulent.

23. Involucre 3.0–4.5 mm high; cypselae strigose; lower leaves less than seven times longer than wide . . . . 28. *S. nemoralis*

23. Involucre 4.5–6.0 (–6.5) mm high; cypselae sericeous; lower leaves more than seven times longer than wide . . . . 29. *S. decemflora*

12. Stems glabrous, at least below the inflorescence.

24. Branches of the inflorescence glabrous.

25. Leaves distinctly 3-nerved above the base; cypselae glabrous or sparsely hispid . . . . 20. *S. glaberrima*

25. Leaves 1-nerved (sometimes obscurely 3-nerved); cypselae puberulent throughout . . . . 19. *S. juncea*

24. Branches of the inflorescence pubescent.

26. Leaves somewhat fleshy, usually entire . . . . 16. *S. sempervirens*

26. Leaves not fleshy, some or all of them serrate.

27. Basal leaves absent at flowering time.

28. Leaves 3-veined above the base; stems usually strongly glaucous . . . . 27. *S. gigantea*

28. Leaves 1-veined; stems not glaucous.

29. Lower leaves larger than the middle and upper leaves . . . . 19. *S. juncea*

29. Lower leaves smaller than the middle and upper leaves . . . . 21. *S. ulmifolia*

27. Basal leaves present at flowering time.

30. Stems strongly angular, sometimes narrowly winged; leaves harshly scabrous above . . . . 14. *S. patula*

30. Stems terete, unwinged; leaves glabrous or spreading-pubescent, not harshly scabrous above.

31. Leaves sharply serrate; lower leaves with winged petioles; upper and middle leaves more than 2 cm wide . . . . 12. *S. arguta*

31. Leaves sparsely serrate to often entire; petioles of lower leaves unwinged; upper and middle leaves less than 2 cm wide.

32. Leaves sometimes slightly sweet-scented; heads not secund; involucre 5–6 mm high . . . . 8. *S. jejunifolia*

32. Leaves not sweet-scented; heads secund; involucre 3–5 mm high . . . . 17. *S. uliginosa*

1. **Solidago petiolaris** Ait. Hort. Kew. 3:216. 1789. Fig. 12.

Perennial from a thickened caudex and usually long creeping rhizomes; stems 1 to several, erect, to 1.5 m tall, spreading-pubescent, at least above, often glabrous below; leaves firm, the basal absent at flowering time, elliptic to obovate, acute at the apex, tapering to the sessile or short-petiolate base, usually sparingly serrate, glabrous above, not glutinous, pubescent below, 1-veined, 6–12 cm long, 2.5–6.0 cm wide; middle and upper leaves narrowly elliptic to narrowly oblong, acute at the apex, tapering or rounded at the sessile base, entire or sparsely denticulate, densely spreading-pubescent, usually scabrous, not glutinous; inflorescence of short,

12. *Solidago petiolaris* (Downy goldenrod).
a. Lower part of plant.
b. Upper part of plant.
c. Flowering head.
d. Phyllaries.
e. Ray flower.
f. Disc flower.
g. Cypsela.

leafy-bracted, axillary racemes or a narrow panicle; involucre 4.5–7.5 mm high; phyllaries in 3–5 series, unequal, oblong-lanceolate, with acute and often squarrose tips, glabrous or sparsely pubescent, not glandular, not viscid; receptacle epaleate; ray flowers 5–8, 3.5–7.0 mm long, pistillate, yellow; disc flowers 4–12, 3–5 mm high, bisexual, yellow; cypselae flat, obovoid, 3–4 mm long, glabrous or nearly so; pappus of capillary bristles 2–4 mm long.

Common Name: Downy goldenrod.
Habitat: Bluff tops, dry woods.
Range: North Carolina to Illinois to Nebraska, south to New Mexico and Florida.
Illinois Distribution: Confined to the southern one-sixth of the state.

Plants with stems without spreading hairs, with glutinous leaves not more than 3 cm wide, and with glandular-hairy, viscid phyllaries are considered in this work to be a different species known as *S. angusta*. Nesom in 1990 studied the *S. petiolaris* complex and concluded that the range of variation in *S. petiolaris* did not justify dividing it into other taxa. After observing the *S. petiolaris* complex in the field for more than sixty years, I am able to distinguish *S. petiolaris* from *S. angusta* in most cases. I have just as much trouble distinguishing some specimens of *S. petiolaris* from *S. buckleyi*.

Solidago buckleyi is a species very similar to *S. petiolaris*, differing by its wider, thinner leaves that are more sharply toothed and by its usually slightly shorter cypselae.

The flowers of *S. petiolaris* bloom from early June to October.

2. **Solidago angusta** (Torr. & Gray) Gray, Fl. N. Am. 2:204. 1842. Fig. 13.
Solidago petiolaris Ait. var. *angusta* (Torr. & Gray) Gray, Proc. Am. Acad. Arts 17:189. 1882.
Solidago wardii Britt. Man. Fl. N. States & Can. 935. 1901.
Solidago petiolaris Ait. var. *wardii* (Britt.) Fern. Rhodora 10:87. 1908.

Perennial from a thickened caudex and usually long creeping rhizomes; stems 1 to several, erect, to 1.5 m tall, pubescent at least above but the hairs not spreading, sometimes glabrous below; leaves firm, the basal absent at flowering time, narrowly elliptic, acute at the apex, tapering to the sessile or short-petiolate base, sparsely serrate to entire, often glabrous above, pubescent and glutinous below, 1-nerved, 5–10 cm long, 1–3 cm wide; middle and upper leaves linear, acute at the apex, tapering or rounded at the sessile base, usually entire, pubescent, glutinous; inflorescence of short, leafy-bracted, axillary racemes or a narrow raceme-like panicle; involucre 4.0–6.5 mm high; phyllaries in 3–5 series, unequal, linear to narrowly lanceolate, acute at the apex and often with squarrose tips, glandular-pubescent, viscid; receptacle epaleate; ray flowers 5–7, 3–6 mm long, pistillate, yellow; disc flowers 4–10, 3.0–4.5 mm high, bisexual, yellow; cypselae 3–4 mm long, narrowly obovoid, flat, glabrous or nearly so; pappus of capillary bristles 2–4 mm long.

Common Name: Sticky goldenrod.
Habitat: Rocky woods, bluff tops.
Range: Illinois to Kansas, south to Texas and Louisiana.
Illinois Distribution: Apparently confined to the southern one-sixth of the state.

Most botanists equate this species with *S. petiolaris*, and intergradation does occur, but *S. angusta* typically has narrower leaves not more than 3 cm wide; viscid, spreading hairs on the stem and leaves; and narrower, glandular-hairy, viscid phyllaries.

Solidago angusta flowers from early June through October.

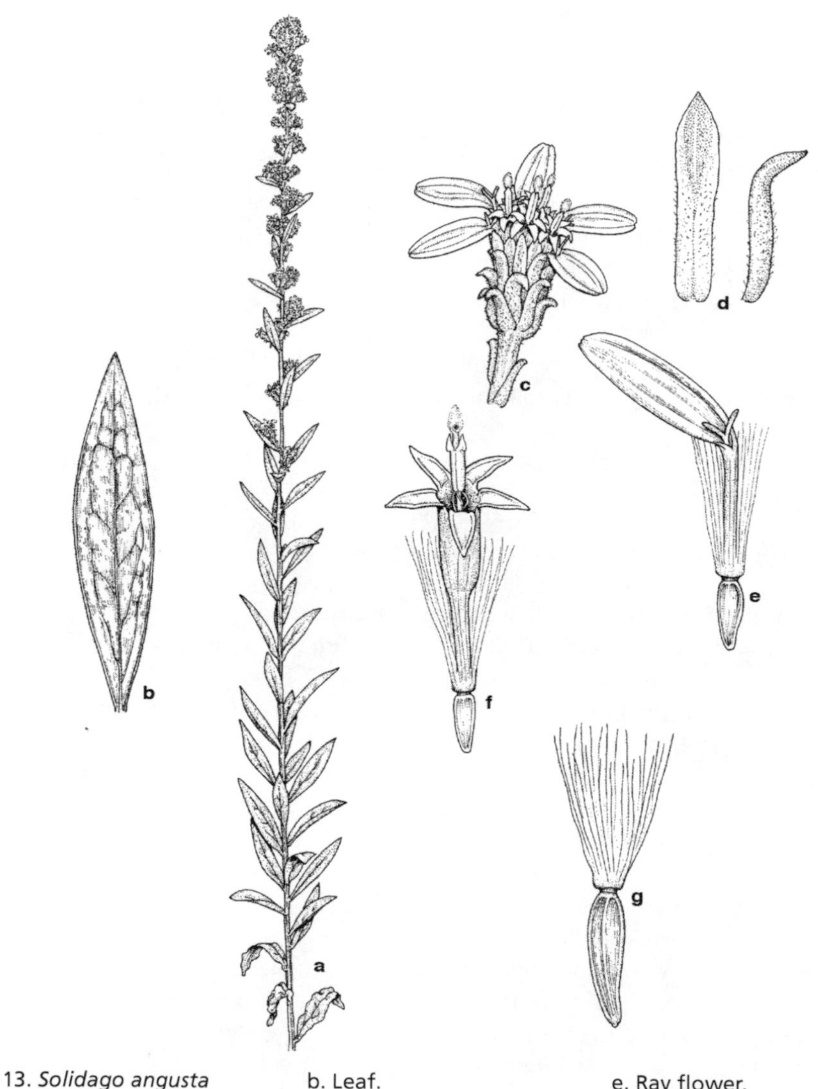

13. *Solidago angusta* (Sticky goldenrod).
a. Upper part of plant.
b. Leaf.
c. Flowering head.
d. Phyllaries.
e. Ray flower.
f. Disc flower.
g. Cypsela.

3. **Solidago buckleyi** Torr. & Gray, Fl. N. Am. 2:198. 1842. Fig. 14.

Perennial from a stout caudex, without rhizomes; stems 1 to several, erect, to 1.5 m tall, short-pilose or glabrous above, often glabrous near the base, longitudinally striate; basal leaves absent at flowering time; cauline leaves thin, uniform in shape, the middle ones larger than the lower ones, elliptic to obovate, acute at the apex, tapering to the sessile or short-petiolate base, sharply serrate, glabrous or nearly so to short-pilose on the veins on the upper surface, hispid on the lower surface,

14. *Solidago buckleyi* (Buckley's goldenrod).

a. Habit.
b. Back side of leaf.
c. Flowering head.
d. Ray flower.
e. Disc flower, side view.
f. Disc flower, face view.

particularly on the veins, 1-nerved, to 15 cm long, to 5 cm wide; inflorescence of axillary clusters or a narrow raceme-like panicle, with short branchlets, the heads not secund; involucre 4–6 (–8) mm high; phyllaries in 3 or 4 series, unequal, the outer phyllaries acute to acuminate, spreading or reflexed at the tip, glandular-pubescent, the margins ciliate, the inner phyllaries obtuse, not spreading; receptacle epaleate; ray flowers 5–7 (–9), 3–5 mm long, pistillate, yellow; disc flowers up to 15, 3–5 mm high, bisexual, yellow; cypselae flat, obovoid, glabrous or nearly so, 2.0–2.7 mm long; pappus of capillary bristles 4–5 mm long.

Common Name: Buckley's goldenrod.
Habitat: Bluff tops, dry forests, rocky woods.
Range: Arkansas, Illinois, Indiana, Kentucky, and Missouri.
Illinois Distribution: Occasional in the southern one-sixth of the state, but extending north to St. Clair County.

This species strongly resembles *S. petiolaris* but differs by its broader, more sharply serrate leaves, hairier lower leaf surfaces, and shorter cypselae. It often grows in the vicinity of *S. petiolaris*. *Solidago buckleyi* also differs from the somewhat similar *S. hispida* by its cauline leaves that are longer and wider than the basal leaves.

This species flowers from August to October.

4. **Solidago bicolor** L. Syst. Nat. ed. 12, 2:556. 1767. Fig. 15.

Perennial from a stout caudex, without rhizomes; stems 1 to a few, erect, to 1.0 (–1.2) m tall, with spreading pubescence; basal leaves usually persistent, rather firm, unequal in size, the basal larger than the middle cauline leaves, broadly lanceolate to oblanceolate, acute at the apex, tapering or less commonly abruptly contracted to a petiole, serrate to crenate, with spreading pubescence on both surfaces, to 20 cm long, to 5 cm wide; upper leaves progressively smaller, sessile, entire; inflorescence narrow and elongated or sometimes in axillary clusters, the heads not secund; involucre 3–5 (–6) mm high; phyllaries in 3 or 4 series, unequal, linear-lanceolate to oblong, obtuse, glabrous, green-tipped but with a white margin; receptacle epaleate; ray flowers 7–9, 2–4 mm long, pistillate, silvery white to cream; disc flowers up to 12, 3–4 mm high, bisexual, yellowish; cypselae flat, obovoid, glabrous or sparsely strigose, 1.5–2.5 mm long; pappus of capillary bristles 2.5–3.5 mm long, often clavate.

Common Name: Silverrod; white goldenrod.
Habitat: Dry open woods.
Range: Nova Scotia to Wisconsin, south to Missouri, Mississippi, and Georgia.
Illinois Distribution: Scattered in Illinois but not common.

This species is similar to *S. hispida* but differs by its silvery white or cream ray flowers. The pappus is more clavate than in *S. hispida*.

Solidago bicolor flowers from August to October.

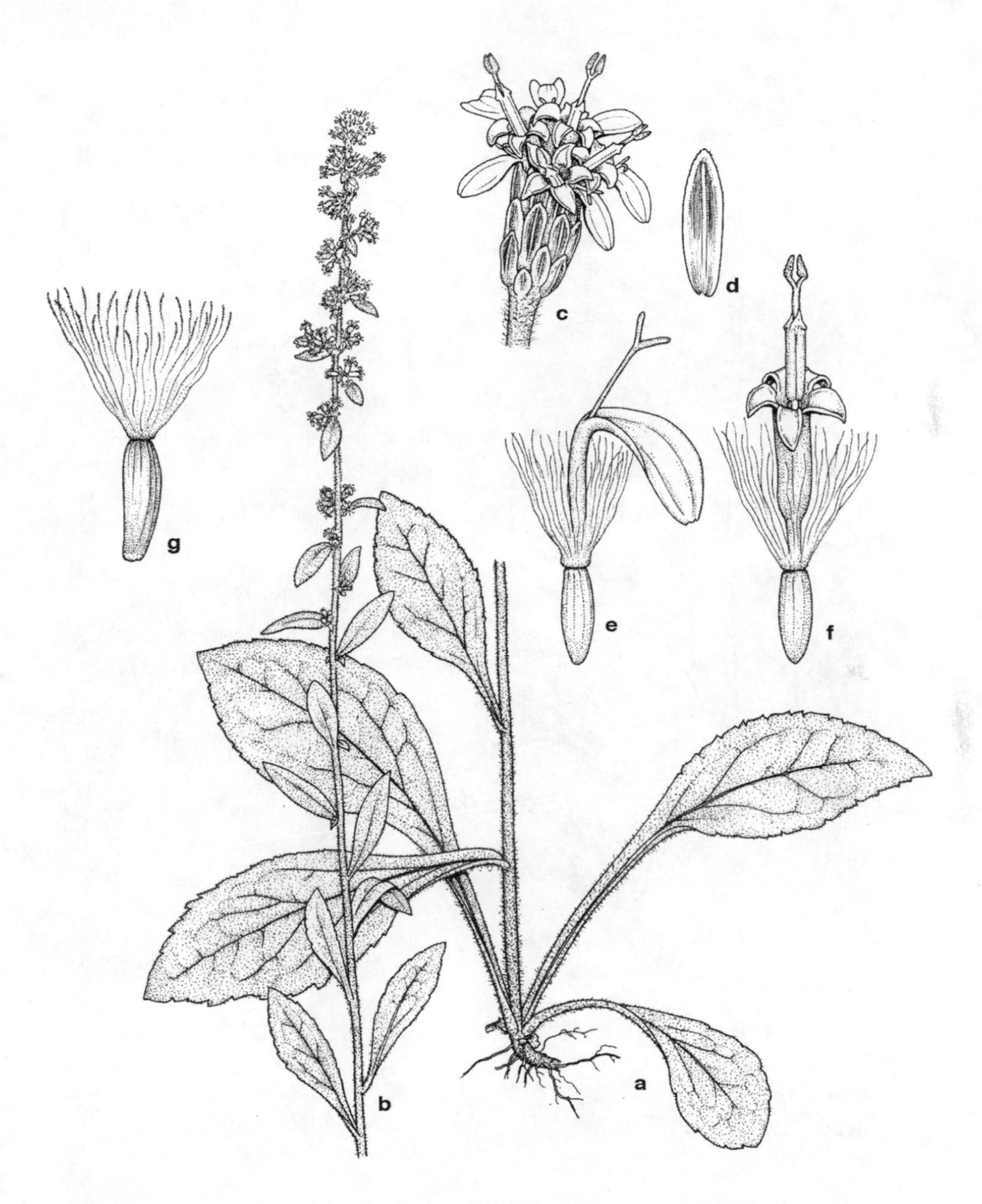

15. *Solidago bicolor* (Silverrod).
a. Lower part of plant.
b. Upper part of plant.
c. Flowering head.
d. Phyllary.
e. Ray flower.
f. Disc flower.
g. Cypsela.

16. *Solidago hispida* (Hairy goldenrod).
a. Lower part of plant.
b. Upper part of plant.
c. Flowering head.
d. Phyllary.
e. Ray flower.
f. Disc flower.
g. Cypsela.

5. **Solidago hispida** Muhl. ex Willd. Sp. Pl. 3:2063. 1803. Fig. 16.
Solidago lanata Hook. Fl. Bor. Am. 2:4. 1834.
Solidago bicolor L. var. *concolor* Torr. & Gray, Fl. N. Am. 197. 1842.
Solidago bicolor L. var. *lanata* (Hook.) Gray, Proc. Am. Acad. Art. 17:190. 1892.
Solidago bicolor L. var. *hispida* (Muhl. ex Willd.) BSP. Prelim Cat. 20. 1898.
Solidago hispida Muhl. ex Willd. var. *lanata* (Hook.) Fern. Rhodora 10:87. 1908.

Perennial from a stout caudex, without rhizomes; stem solitary, erect, to 1 m tall, spreading pilose, rarely nearly glabrous or rarely densely villous, longitudinally striate; basal leaves persistent at flowering time, firm, larger than the cauline leaves, obovate to elliptic, acute at the apex, tapering to a long, winged petiole, shallowly serrate or crenate, usually densely pubescent on both surfaces, 1-veined, to 20 cm long, to 5 cm wide; cauline leaves similar but progressively smaller, often entire, sessile; inflorescence in short racemes, the racemes often axillary but occasionally terminal, forming a narrow panicle, the heads not secund; involucre (3.0–) 3.5–6.0 mm high; phyllaries in 3–5 series, unequal, linear-oblong, obtuse to acute, appressed, glabrous and sometimes ciliate; receptacle epaleate; ray flowers 7–15, 4–6 mm long, pistillate, orange-yellow; disc flowers up to 15, 2.5–5.0 mm high, bisexual, orange-yellow; cypselae flat, obovoid, glabrous or sometimes sparsely pubescent when immature, 0.8–1.5 mm long; pappus of capillary bristles 2.5–3.0 mm long.

Common Name: Hairy goldenrod.
Habitat: Dry open woods.
Range: Newfoundland to Saskatchewan, south to Oklahoma, Louisiana, and Georgia.
Illinois Distribution: Jackson, Randolph, and Union counties.

This mostly Ozarkian species is distinguished by its axillary flowering clusters or narrow panicles, its spreading hairy stems, and its firm, persistent basal leaves.

A specimen from Cedar Lake Recreation Area in Jackson County, collected by Thomas Heineke, has densely villous stems. It may be known as var. *lanata* (Hook.) Fern.

Solidago hispida flowers from August to October.

6. **Solidago speciosa** Nutt. Gen. 2:160. 1818. Fig. 17.
Solidago emarginata Millsp. & Sherff, Field Mus. Publ. Bot. 4:7. 1918.

Perennial from a thickened caudex, without rhizomes; stems 1 to several, erect, to 2 m tall, glabrous, sometimes scabrous, sometimes puberulent on the branches of the inflorescence, with longitudinal striations; lowest leaves persistent, firm, oblanceolate to ovate to elliptic, acute at the apex, long-petiolate at the base, glabrous or nearly so, serrate, to 30 cm long, 3.5–10.0 cm wide; cauline leaves firm, 20 or more, oblong to narrowly lanceolate, acute at the apex, tapering to the sessile base, usually entire, more than 2.5 cm wide; inflorescence a dense panicle with crowded ascending branches, the numerous heads not secund; involucre 3–6 mm high;

17. *Solidago speciosa* (Showy goldenrod).
a. Lower part of plant.
b. Upper part of plant.
c. Flowering head.
d. Phyllary.
e. Ray flower.
f. Disc flower.

phyllaries in 3–5 series, unequal, obtuse, glabrous or with ciliate margins, sometimes glutinous; receptacle epaleate; ray flowers 5–8, 3–5 mm long, pistillate, yellow; disc flowers (4–) 7–10, 3–5 mm high, bisexual, yellow; cypselae flat, obovoid, 1.0–1.8 mm long, glabrous; pappus of capillary bristles 2.0–3.5 mm long.

Common Name: Showy goldenrod.
Habitat: Dry woods, prairies, black oak savannas.
Range: Massachusetts to Minnesota, south to Arkansas, Tennessee, and North Carolina.
Illinois Distribution: Scattered throughout the state, but less common in the southern tip of the state.

Three species are in the *S. speciosa* complex in Illinois, although most botanists consider them as either varieties of *S. speciosa* or not worthy of recognition at all. After observing this complex in Illinois and adjacent states for more than sixty years, I usually can readily distinguish the three types. In this work I am recognizing them as three separate species.

As treated here, *S. speciosa* is a species with all leaves more than 2.5 cm wide, with the basal leaves serrate and persistent at flowering time, and with more than 20 cauline leaves. *Solidago rigidiuscula* is similar but with the basal leaves withered by flowering time, usually entire, and the leaves much more narrow. The similar *S. jejunifolia* has a more open, less dense panicle than either *S. speciosa* or *S. rigidiuscula*, and has fewer than 20 leaves on the stem; *Solidago jejunifolia* often has slightly sweet-scented leaves, particularly in the morning.

Millspaugh and Sherff's *Solidago emarginata*, from near Morris, is a synonym for *S. speciosa*.

Solidago speciosa flowers from August to October.

7. **Solidago rigidiuscula** (Torr. & Gray) Porter, Mem. Torrey Club 5:319. 1894. Fig. 18.
Solidago speciosa Nutt. var. *rigidiuscula* Torr. & Gray, Fl. N. Am. 2:205. 1841.

Perennial from a thickened caudex, without rhizomes; stems 1 to several, erect, to 1 m tall, glabrous, sometimes scabrous, sometimes puberulent in the inflorescence, with longitudinal striations; lowest leaves withered at flowering time, firm, narrowly lanceolate to narrowly oblong, acute at the apex, long-petiolate at the base, usually serrate, glabrous or nearly so, to 12 cm long, to 2.5 cm wide; cauline leaves firm, 20 or more, narrowly oblong, acute at the apex, tapering to the sessile base, entire, often somewhat scabrous, to 8 cm long, less than 2.5 cm wide; inflorescence a dense panicle, with crowded ascending branches, the numerous heads not secund; involucre 3–6 mm high; phyllaries in 3–5 series, unequal, oblong, usually obtuse, glabrous or with ciliate margins, often glutinous; receptacle epaleate; ray flowers 5–8, 3–5 mm long, pistillate, yellow; disc flowers 4–10, 3–5 mm high, bisexual, yellow; cypselae flat, obovoid, glabrous, 1–2 mm long; pappus with capillary bristles 2.0–3.5 mm long.

Common Name: Prairie goldenrod.
Habitat: Prairies, dry woods.
Range: Ohio to Saskatchewan, south to Colorado, Texas, Louisiana, and Alabama.
Illinois Distribution: Scattered throughout the state.

Solidago rigidiuscula is similar to *S. speciosa* but is more slender, its basal leaves are withered at flowering time, and the leaves are much narrower, never exceeding 2.5 cm.
This species flowers from August to October.

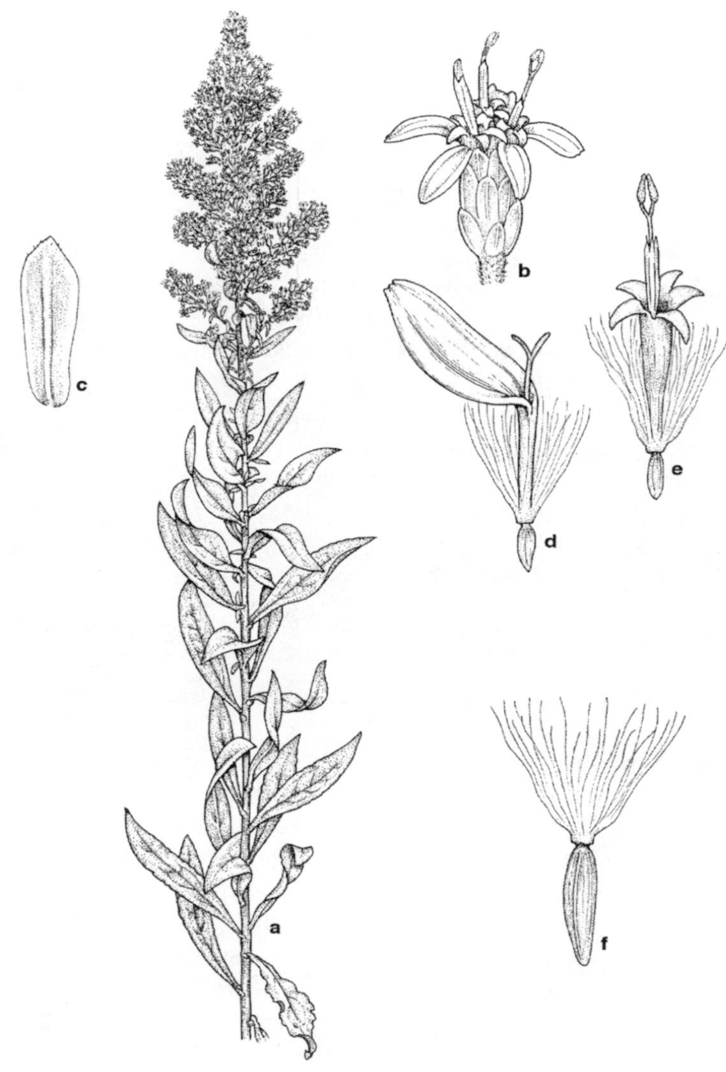

18. *Solidago rigidiuscula* (Prairie goldenrod).
a. Upper part of plant.
b. Flowering head.
c. Phyllary.
d. Ray flower.
e. Disc flower.
f. Cypsela.

8. **Solidago jejunifolia** Steele, Contr. U. S. Natl. Herb. 16:223. 1913. Fig. 19.
Solidago speciosa Nutt. var. *jejunifolia* (Steele) Cronq. Rhodora 49:77. 1947.

Perennial from a thickened caudex, without rhizomes; stems 1 to several, erect, to 0.8 m tall, glabrous, usually hispidulous on the branches of the inflorescence, with longitudinal striations; lower leaves persistent at flowering time, faintly sweet-scented, thick, firm, ovate to oblong, acute or obtuse and apiculate at the apex, long-petiolate at the base, entire or shallowly serrate, glabrous, to 17 cm long, less

19. *Solidago jejunifolia* (Few-leaved goldenrod).
a. Lower part of stem.
b. Upper part of stem.
c. Flowering head.
d. Phyllary.
e. Ray flower.
f. Disc flower.
g. Cypsela.

than 1 cm wide; cauline leaves fewer than 20, less than 2 cm wide, lanceolate to linear, smaller than the basal leaves, entire, glabrous; inflorescence a loose and open panicle with few branches, the heads not secund; involucre 5–6 mm high; phyllaries in 3 or 4 series, unequal, oblong to linear-oblong, usually obtuse, glabrous; receptacle epaleate; ray flowers 5–9, 3–5 mm long, pistillate, yellow; disc flowers 5–7, 3–5 mm high, bisexual, yellow; cypselae flat, obovoid, glabrous, 1–2 mm long; pappus of capillary bristles 2.0–3.5 mm long.

Common Name: Few-leaved goldenrod.
Habitat: Wet, sandy soil.
Range: Ontario south to northern Illinois.
Illinois Distribution: Lake and McHenry counties.

This species, often included within *S. speciosa*, is distinguished by its fewer, narrower cauline leaves that are faintly sweet-scented, by its loose, open inflorescence with fewer branches, and by its longer involucres. The leaves have a faintly sweet odor, particularly noticeable early in the morning.

Solidago jejunifolia flowers in August and September.

9. **Solidago sciaphila** E. S. Steele, Contr. U.S. Natl. Herb. 13:371–372. 1911. Fig. 20.

Perennial from a thickened caudex; stems 1 to several, erect, to 1 m tall, glabrous except for puberulent inflorescence branches; leaves thin, almost semisucculent, the lowest persistent at flowering time, broadly elliptic to obovate, acute at the apex, tapering to the petiolate base, crenate to serrate, glabrous, to 15 cm long, to 8 cm wide; middle and upper leaves smaller, often sessile and often entire; inflorescence of short, axillary racemes, forming a thyrse, the heads not secund; involucre 4–7 mm high; phyllaries in 3 or 4 series, unequal, oblong, obtuse at the apex, glabrous; receptacle epaleate; ray flowers 5–10, usually 7 or 8, 2.5–4.0 mm long, pistillate, yellow; disc flowers up to 10, 3–4 mm high, bisexual, yellow; cypselae flat, obovoid, strigose, 2.5–3.0 mm long; pappus of capillary bristles 2.5–4.0 mm long.

Common Name: Shadowy goldenrod.
Habitat: Limestone and sandstone cliffs.
Range: Illinois, Iowa, Minnesota, Wisconsin.
Illinois Distribution: Carroll, Jo Daviess, LaSalle, and Ogle counties.

This species is distinguished by its glabrous stems and inflorescence of short axillary racemes forming a thyrse.

This species flowers in August and September.

20. *Solidago sciaphila* (Shadowy goldenrod).
a. Habit.
b. Flowering head.
c. Phyllary.
d. Ray flower.
e. Disc flower.
f. Cypsela.
g. Basal rosette of leaves.

10. **Solidago caesia** L. Sp. Pl. 879. 1753. Fig. 21.

Perennial from a thickened caudex and usually long creeping rhizomes; stems 1 to several, erect to arching, to 1 m tall, glabrous, strongly glaucous; basal leaves absent at flowering time, thin, elliptic-lanceolate to elliptic, acuminate at the apex, tapering to the sessile base, sharply serrate, glabrous on both surfaces, 1-veined, to 10 cm long, to 3 cm wide; middle and upper leaves lanceolate to narrowly oblong, acuminate at the apex, tapering to the sessile base, sharply serrate, glabrous on both surfaces or rarely with pubescence on the veins on the lower surface, to 8 cm long, to 2 cm wide; inflorescence axillary, in short, leafy clusters, the heads not secund; involucre 2.5–5.0 mm high; phyllaries in 3–5 series, unequal, linear-lanceolate to oblong, obtuse or acute at the apex, appressed, glabrous with ciliate margins; receptacle epaleate; ray flowers (1–)2–5, 3.0–4.5 mm long, pistillate, yellow; disc flowers 3–8, 3–4 mm high, bisexual, yellow; cypselae flat, obovoid, pubescent, 1.8–2.6 mm long; pappus of capillary bristles 2.5–3.0 mm long.

21. *Solidago caesia* (Blue-stemmed goldenrod). a. Upper part of plant. b. Flowering head. c. Ray flower. d. Disc flower.

Common Name: Blue-stem goldenrod; wreath goldenrod.
Habitat: Mesic woods, bottomland forests, upland woods, black oak savannas, along rivers and streams.
Range: Quebec to Wisconsin, south to Texas and Florida.
Illinois Distribution: Occasional throughout the state.

The extremely glaucous leaves and stems are distinctive and account for the common name of blue-stem goldenrod. The short axillary racemes of flowers are also distinctive and give a very attractive appearance to this species.

Solidago caesia flowers from August to October.

11. **Solidago flexicaulis** L. Sp. Pl. 879. 1753. Fig. 22.
Solidago latifolia L. Sp. Pl. 879. 1753.

Perennial from long creeping rhizomes; stems usually solitary, slender, erect, more or less zigzag, to 1 m tall, glabrous except in the inflorescence, with a few ridges;

22. *Solidago flexicaulis* (Zigzag goldenrod).
a. Upper part of plant.
b. Leaf.
c. Flowering head.
d. Phyllary.
e. Ray flower.
f. Disc flower.
g. Cypsela.

basal leaves absent at flowering time, thin, broadly ovate to elliptic-ovate, acute at the apex, rounded at the base, sharply serrate, glabrous above, pubescent below, 1-veined, to 15 cm long, to 10 cm wide, the petioles winged; middle cauline leaves thin, ovate-lanceolate to ovate, acuminate at the apex, rounded at the base, sharply serrate, glabrous or with appressed pubescence on the upper surface, pilose or hirsute, at least on the veins on the lower surface, to 10 cm long, to 6 cm wide, the petiole winged; inflorescence of numerous heads in short racemes, some of the racemes axillary, the heads not secund; involucre 4–6 mm high; phyllaries in 3–5 series, unequal, linear-oblong to oblong, obtuse or acute at the apex, appressed, glabrous except for the sometimes ciliate margin; receptacle epaleate; ray flowers 3–5, 3–5 mm long, pistillate, yellow; disc flowers 5–12, 4–7 mm high, bisexual, yellow; cypselae flat, obovoid, puberulent, 1.5–2.5 mm long; pappus of capillary bristles 2–3 mm long.

Common Name: Zigzag goldenrod; broad-leaved goldenrod.
Habitat: Mesic woods, bottomland forests, calcareous springy places, calcareous slopes.
Range: New Brunswick to Minnesota, south to Arkansas, Alabama, and Georgia.
Illinois Distribution: Occasional throughout the state.

The stems usually have a zigzag appearance, particularly as the season progresses. The leaves are also distinctive because of their mostly ovate shape with very sharp teeth.

The flowers bloom from August to October.

12. **Solidago arguta** Ait. Hort. Kew. 3:213. 1789. Fig. 23.
Solidago arguta Ait. var. *caroliniana* Gray, Syn. Fl. N. Am. 1:155. 1884.
Solidago arguta Ait. ssp. *caroliniana* (Gray) C. H. Morton, Phytologia 28:1. 1974.

Perennial from a stout caudex; stems 1 to several, erect, to 1.2 m tall, glabrous or sometimes with pubescence below, longitudinally striate; leaves rather thin, the lowest ones ovate to broadly elliptic, persistent at flowering time, acute at the apex, tapering to a winged petiole, to 30 cm long, to 12 cm wide, sharply toothed, glabrous or scabrous above, usually with a few short spreading hairs below, or sometimes glabrous; middle and upper leaves elliptic to lanceolate, acute at the apex, tapering at base to a very short petiole or sometimes sessile, to 10 cm long; inflorescence an open pyramidal panicle with numerous secund heads on recurved branches, the branches more or less pilose; involucre 3.5–6.0 mm high; bracts in 3–5 series, unequal, about 1 mm wide, not keeled, linear-oblong, acute or obtuse at the apex, appressed, glabrous or nearly so except ciliate whitish margins; receptacle epaleate; ray flowers (2–)5–8, 4–6 (–8) mm long, pistillate, yellow; disc flowers 8 or more, 3–4 mm high, bisexual, yellow; achenes flat, sparsely pubescent or glabrous, narrowly obovoid, 1.5–2.5 mm long; pappus of capillary bristles 2.5–3.5 mm long.

Common Name: Sharp-toothed goldenrod.
Habitat: Dry forests, rocky slopes.

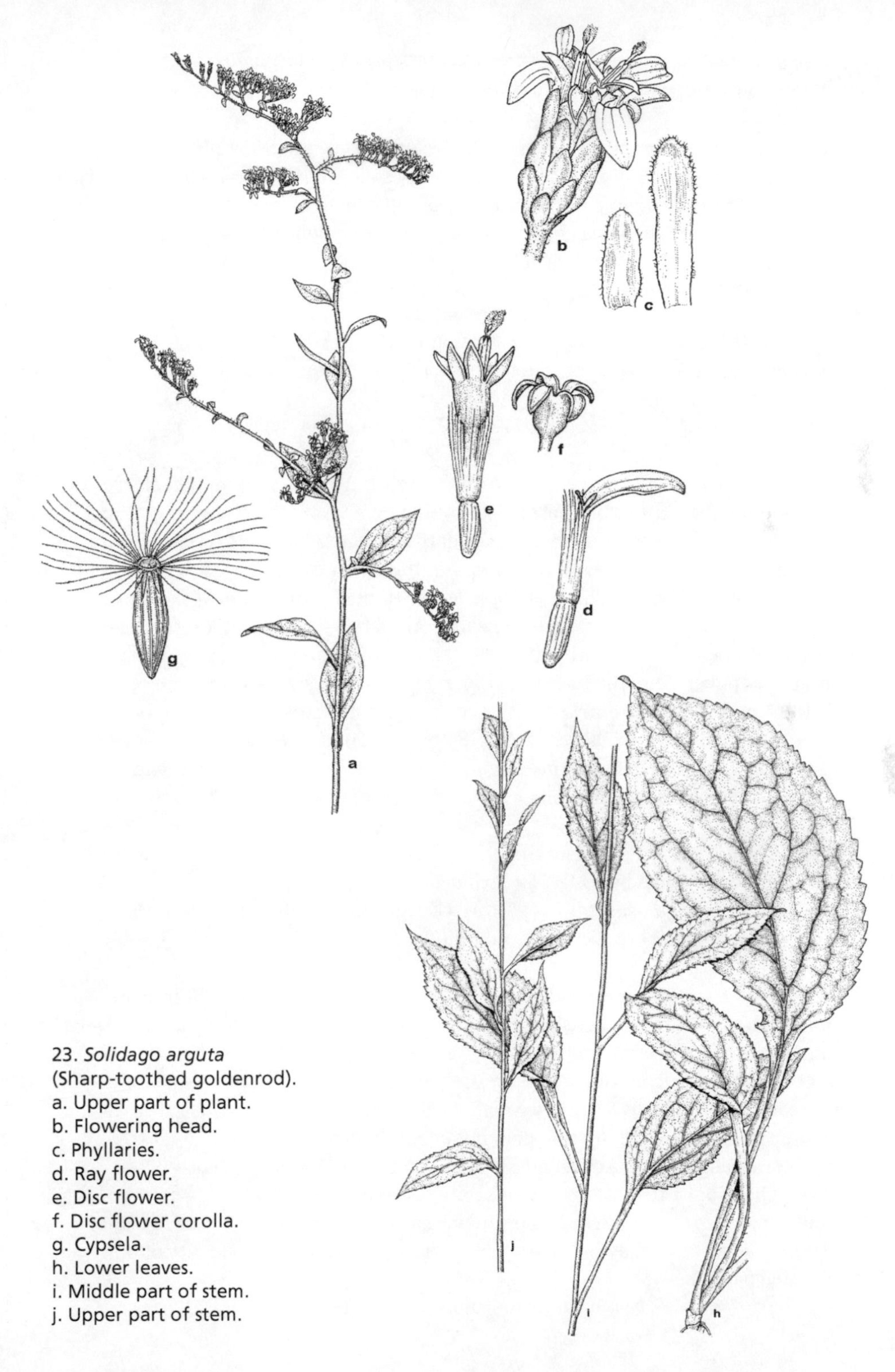

23. *Solidago arguta*
(Sharp-toothed goldenrod).
a. Upper part of plant.
b. Flowering head.
c. Phyllaries.
d. Ray flower.
e. Disc flower.
f. Disc flower corolla.
g. Cypsela.
h. Lower leaves.
i. Middle part of stem.
j. Upper part of stem.

Range: Maine to Ontario, south to Missouri, Tennessee, and North Carolina.
Illinois Distribution: Known only from Jackson and Union counties.

Rare plants with completely glabrous leaves and stems may be known as var. *caroliniana*. It apparently has been erroneously called *S. boottii*, which is an entirely different species with more pubescent cypselae, pubescent leaves, and moderately hairy stems. Var. *caroliniana* has been found on rocky slopes beneath *Pinus echinata* in the Pine Hills of Union County.

The flowers bloom in August and September.

13. **Solidago boottii** Hook. Hook. Compan. Bot. Mag. 1:97. 1833. Fig. 24.
Solidago arguta Ait. var. *boottii* (Hook.) Palmer & Steyerm. Ann. Mo. Bot. Gard. 22:659. 1935.
Solidago arguta Ait. ssp. *boottii* (Hook.) G. H. Morton, Phytologia 28:1. 1974.

Perennial from a stout caudex; stems 1-several, erect, to 1.5 mm tall, glabrous above but pubescent below; basal leaves rather firm, ovate to broadly elliptic, persistent at flowering time, acute to acuminate at the apex, abruptly contracted to a winged petiole, sharply serrate, glabrous on the upper surface, densely pubescent on the lower surface; middle and upper leaves elliptic, acute at the apex, tapering to the nearly sessile base, serrate to nearly entire, to 8 cm long; inflorescence an open pyramidal panicle with numerous secund heads on recurving branches, the branches puberulent; involucre 3–6 mm high; phyllaries in 3–5 series, unequal, linear-oblong, about 1 mm wide, not keeled, acute or obtuse, glabrous or nearly so; receptacle epaleate; ray flowers 2–8, 5–8 mm long, pistillate, yellow; disc flowers 6 or more, 2.8–3.5 mm high, bisexual, yellow; cypselae flat, pubescent with short hairs, obovoid, 2.0–2.8 mm long; pappus of capillary bristles 2.5–3.5 mm long.

Common Name: Boott's goldenrod.
Habitat: Rocky slopes beneath *Pinus echinata*.
Range: Illinois to Kansas, south to Texas, Georgia, and South Carolina.
Illinois Distribution: Known only from Union County.

This plant is sometimes considered to be a variety of *S. arguta* but differs by its pubescent cypselae that are slightly longer than the cypselae of *S. arguta* and by its much more pubescent lower leaf surfaces.

One specimen from Union County with leaves pilose on the lower surface I erroneously identified as *S. strigosa* Small. *Solidago strigosa* is actually a species of the southeastern United States that does not occur in Illinois.

One specimen from a cherty slope in Union County has lower leaves pilose throughout but with the inner bracts only 0.5 mm wide, cypselae 1.3–1.5 mm long, and pappus up to 2.5 mm long. This may be a hybrid between *S. arguta* Ait. and *S. ulmifolia* L., known as **S. X neurolepis** Fern. It occurred in the vicinity of both reputed parents.

The flowers of *S. boottii* bloom in August and September.

24. *Solidago boottii* (Boott's goldenrod).
a. Lower leaf.
b. Middle part of stem.
c. Upper part of plant.
d. Portion of upper part of stem.
e. Flowering head.
f. Phyllary.
g. Ray flower.
h. Disc flower.

25. *Solidago patula* (Rough-leaved goldenrod).
a. Lower part of plant.
b. Upper part of plant.
c. Node with petiole.
d. Flowering head.
e. Phyllary.
f. Ray flower.
g. Disc flower.
h. Cypsela.
i. Basal rosette of leaves.
j. Portion of leaf margin with teeth.

14. **Solidago patula** Muhl. ex Willd. Sp. Pl. 3:2059. 1803. Fig. 25.

Perennial from a thickened caudex; stem usually solitary, erect, to 2 m tall, strongly angular and sometimes narrowly winged, with longitudinal striations, usually glabrous or rarely sparsely pubescent, occasionally glaucous; leaves thick, the basal persistent at flowering time, obovate to broadly elliptic, acute to acuminate at the apex, tapering to a long, winged petiole, serrate, harshly scabrous on the upper surface with pustulate-based hairs, usually glabrous or nearly so on the lower surface, with 1 main vein, to 30 cm long, to 10 cm wide; middle and upper leaves abruptly smaller, elliptic to lanceolate, sessile or short-petiolate; inflorescence a broad, open panicle with numerous recurved branches bearing numerous secund heads; involucre 3.0–4.5 mm high; bracts in 3–5 series, unequal, narrowly oblong, obtuse to acute at the apex, appressed, usually glabrous except for the ciliate margins; receptacle epaleate; ray flowers 6–12, 3–7 mm long, pistillate, yellow; disc flowers up to 20, 3–6 mm high, bisexual, yellow; cypselae flat, obovoid, 1.2–1.8 mm long, sparsely and minutely pubescent, rarely glabrous; pappus of capillary bristles 2–3 mm long.

Common Name: Rough-leaved goldenrod; swamp goldenrod.
Habitat: Swampy woods, calcareous fens, bogs, edge of ponds.
Range: Vermont to Wisconsin and Iowa, south to Texas and Florida.
Illinois Distribution: Scattered in the state but absent or rare in the northwestern and extreme southern counties.

This handsome wetland species is readily recognized by its large, harshly scabrous basal leaves on long and winged petioles and its abruptly smaller middle and upper leaves. While most Illinois specimens have sparsely pubescent cypselae, some have glabrous cypselae.

This species flowers from August to October.

15. **Solidago sphacelata** Raf. Ann. Nat. 14. 1820. Fig. 26.
Solidago cordata Short, Trans. Journ. Med. 7:599. 1834.
Brachychaeta sphacelata (Raf.) Britt. in Kearney, Bull. Torrey Club 20:484. 1893.

Perennial from long creeping rhizomes; stems 1 to several, erect, slender, to 1.2 m tall, densely spreading-pubescent or rarely nearly glabrous; basal leaves persistent at flowering time, broadly ovate, acute at the apex, cordate at the base, serrate, to 12 cm long, nearly as broad, pilose on the lower surface, usually glabrous above, on slender petioles; middle and upper leaves smaller, becoming less cordate and finally tapering to the base, minutely pilose, at least on the lower surface; inflorescence a

26. *Solidago sphacelata* (Heart-leaved goldenrod). a. Basal rosette of leaves. b. Middle part of stem. c. Upper part of stem. d. Flowering head. e. Phyllary. f, g. Ray flowers. h. Disc flower.

narrow, spikelike panicle with a few spreading-recurved branches, the heads secund; involucre 3.0–4.5 mm high; phyllaries in 3 or 4 series, unequal, linear-oblong, more or less keeled, obtuse to acute at the apex, glabrous or nearly so; receptacle epaleate; ray flowers 4–6, 2.0–3.5 mm long, pistillate, yellow; disc flowers 4–5, 2–4 mm high, bisexual, yellow; cypselae flat, pubescent, 1.5–2.0 mm long; pappus of extremely short capillary bristles up to 1 mm long, shorter than the cypselae.

Common Name: Heart-leaved goldenrod.
Habitat: River bluffs.
Range: West Virginia and Ohio to Illinois, south to Mississippi and Georgia.
Illinois Distribution: Hardin and Pope counties.

This species is readily recognized by its cordate lower leaves that are persistent at flowering time. It is the only species of *Solidago* in Illinois with pappus shorter than the cypselae. It was at one time segregated into the genus *Brachychaeta*.

This species flowers in August and September.

16. **Solidago sempervirens** L. Sp. Pl. 878. 1753.

Perennial from a short caudex; stems 1-several, erect, to 2 m tall, glabrous or sometimes puberulent in the inflorescence; leaves thick, somewhat fleshy, the basal persistent at flowering time, broadly oblanceolate to oblong, acute at the apex, tapering to the sessile base, serrate or often entire, with or without marginal cilia, glabrous on both surfaces, to 20 cm long, to 5 cm wide; middle and upper leaves smaller and usually entire; inflorescence a panicle or corymb, the lower branches recurved with secund heads; involucre 3.5–7.0 mm high; phyllaries in 3 or 4 series, unequal, linear-oblong, acute to subacute, glabrous; receptacle epaleate; ray flowers 12–17, 4–6 mm long, pistillate, yellow; disc flowers 17–22, 4.0–5.5 mm high, bisexual, yellow; cypselae flat, obovoid, sparsely strigose, 1.0–1.5 mm long; pappus of capillary bristles 3.5–4.0 mm long.

Two varieties occur in Illinois:

a. Leaves not ciliolate, some of them more than 3 cm wide. . . . . 16a. *S. sempervirens* var. *sempervirens*
a. Leaves ciliolate, up to 3 cm wide . . . . 16b. *S. sempervirens* var. *mexicana*

16a. **Solidago sempervirens** L. var. **sempervirens.** Fig. 27.

Leaves not ciliolate, some of them at least 3 cm wide.

Common Name: Seaside goldenrod.
Habitat: Roadsides (in Illinois) where salt has been applied heavily during the winter season.
Range: Newfoundland to Ontario, south to northeastern Illinois, Texas, and Florida.
Illinois Distribution: Confined to the northeastern counties; first collected in Illinois in 1933.

This variety occurs along highways in northeastern Illinois where salt is applied heavily during the winter season, thus emulating conditions found in brackish areas along the southeastern and gulf coasts of the United States. The nonciliate and wider leaves distinguish it from the following variety.

This variety flowers in Illinois in August to November.

16b. **Solidago sempervirens** L. var. **mexicana** (L.) Fern. Rhodora 37:447. 1935.
Solidago mexicana L. Sp. Pl. 2:878. 1753.

Leaves ciliolate, none of them 3 cm wide.

Common Name: Seaside goldenrod.
Habitat: Roadsides (in Illinois) where salt has been applied heavily during the winter season.
Range: Massachusetts to Ontario, south to Texas and Florida.
Illinois Distribution: Confined to the northeastern counties.

This variety differs from var. *sempervirens* by its ciliolate leaves, none of which reaches a width of 3 cm.

Solidago sempervirens var. *mexicana* flowers from August to November.

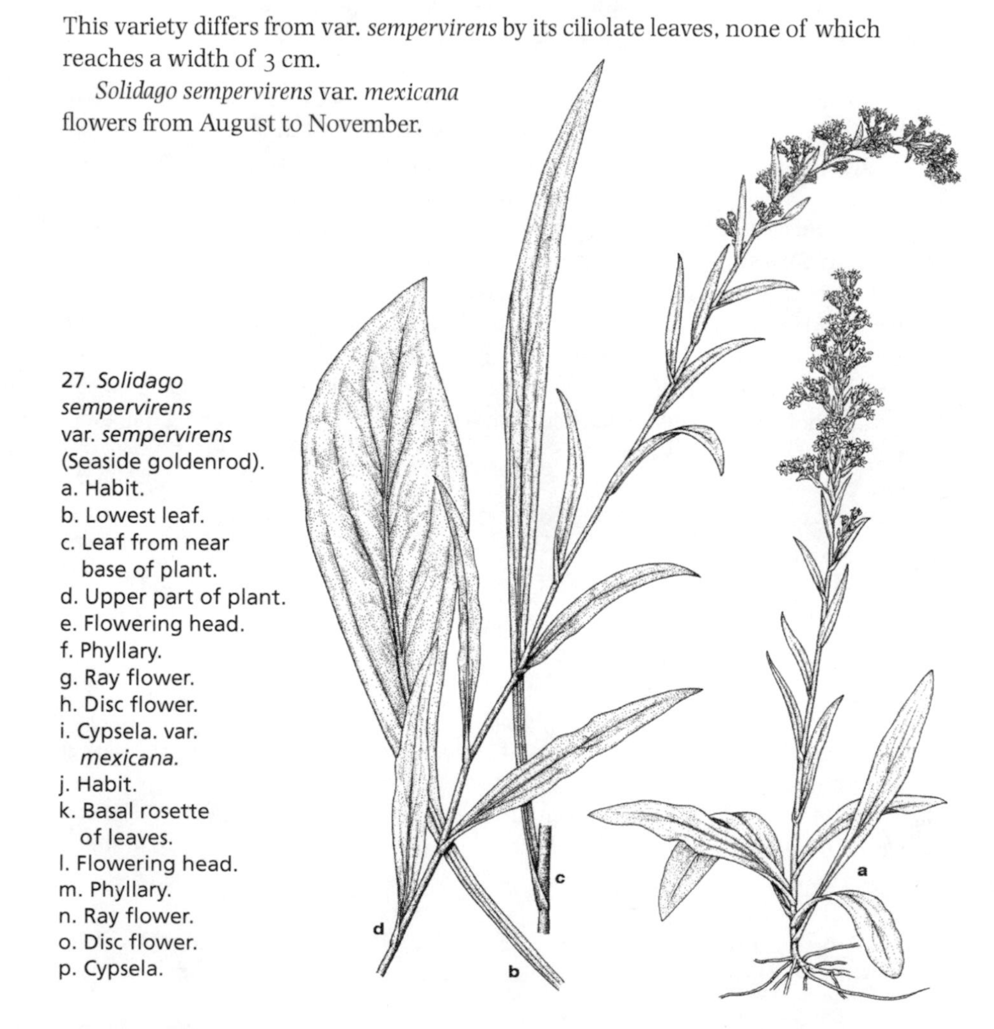

27. *Solidago sempervirens* var. *sempervirens* (Seaside goldenrod).
a. Habit.
b. Lowest leaf.
c. Leaf from near base of plant.
d. Upper part of plant.
e. Flowering head.
f. Phyllary.
g. Ray flower.
h. Disc flower.
i. Cypsela. var. *mexicana.*
j. Habit.
k. Basal rosette of leaves.
l. Flowering head.
m. Phyllary.
n. Ray flower.
o. Disc flower.
p. Cypsela.

f
e
i
h
g
p
j
k
l
m
n
o

17. **Solidago uliginosa** Nutt. Journ. Acad. Nat. Sci. Phila. 7:101. 1834. Fig. 28.
Solidago neglecta Torr. & Gray, Fl. N. Am. 2:213. 1842.

Perennial with a thickened caudex; stem solitary, erect, rather stout, to 1.2 m tall, glabrous or nearly so except in the inflorescence; leaves sometimes slightly succulent, the lowest persistent, lanceolate to oblanceolate, acute or obtuse at the apex, tapering to a long-petiolate base, serrate, glabrous, to 45 cm long, to 8 cm

28. *Solidago uliginosa* (Bog goldenrod).
a. Basal rosette of leaves.
b. Middle part of plant.
c. Upper part of plant.
d. Flowering head.
e. Phyllary.
f. Ray flower.
g. Disc flower.
h. Cypsela.

wide; middle and upper leaves progressively smaller, some of them entire, to 8 cm long; inflorescence a pyramidal panicle with spreading branches to 45 cm long, the heads secund; involucre 3–5 mm high; phyllaries in 3–5 series, unequal, oblong to linear-oblong, obtuse at the apex, glabrous; receptacle epaleate; ray flowers 1–8, 3.2–4.0 mm long, pistillate, yellow; disc flowers 4–8, 4–5 mm high, bisexual, yellow; cypselae flat, obovoid, 1–2 mm long, glabrous or nearly so; pappus of capillary bristles 2.5–3.0 mm long.

Common Name: Bog goldenrod.
Habitat: Acid bogs, calcareous fens.
Range: Nova Scotia to Ontario and Minnesota, south to Illinois and North Carolina.
Illinois Distribution: Northern half of Illinois, south to Cass County.

This species occurs in northern Illinois in acid bogs and calcareous fens. It differs from *S. purshii*, with which it is often combined, by its pyramidal panicle with spreading branches bearing secund heads, its obtuse bracts, and its disc flowers that number 4–8.

The flowers bloom from August to October.

18. **Solidago purshii** Porter, Bull. Torr. Bot. Club 21:311. 1894. Fig. 29.
Solidago humilis Pursh, Fl. Am. Sept. 2:543. 1813, *non* Mill. (1768).

Perennial with a thickened caudex; stem solitary, erect, to 1 m tall, glabrous or nearly so, except in the inflorescence; leaves more or less somewhat succulent, the lowest persistent, lanceolate to oblanceolate, acute or obtuse at the apex, tapering to a long-petiolate base, sparsely serrate or sometimes entire, glabrous, to 45 cm long, to 7 cm wide; middle and upper leaves progressively smaller and often entire, glabrous; inflorescence an elongated thyrse, with strongly ascending branches, the heads not secund; involucre 4–6 mm high; phyllaries in 3–5 series, unequal, lanceolate to oblong, acute to obtuse at the apex, glabrous; receptacle epaleate; ray flowers 4–6, 4–6 mm long, pistillate, yellow; disc flowers 9–15, 4–6 mm high, bisexual, yellow; cypselae flat, obovoid, 1–2 mm long, glabrous or nearly so; pappus of capillary bristles 2.5–3.0 mm long.

Common Name: Pursh's bog goldenrod.
Habitat: Acid bogs.
Range: Labrador to Manitoba, south to Minnesota, Wisconsin, Illinois, Indiana, Ohio, West Virginia, and New York.
Illinois Distribution: Kane, Lake, and McHenry counties.

Although often combined with *S. uliginosa* and sometimes intergrading with it, *S. purshii* in the field may be recognized by its thyrse type of inflorescence and its more numerous disc flowers. In the few Illinois specimens, the phyllaries are usually acute.

The flowers bloom from August to October.

29. *Solidago purshii* (Pursh's bog goldenrod). a. Upper part of plant. b. Leaf. c. Flowering head. d. Phyllaries. e. Ray flower. f. Disc flower. g. Cypsela.

19. **Solidago juncea** Ait. Hort. Kew. 3:213–214. 1789. Fig. 30.
Solidago juncea Ait. f. *scabrella* (Torr. & Gray) Fern. Rhodora 38:408. 1936.

Perennial from a stout caudex and sometimes from short rhizomes; stems 1 to a few, erect, to 1.2 m tall, glabrous throughout, except rarely with hirtellous branches in the inflorescence, longitudinally striate; basal leaves beginning to wither at flowering time, firm, narrowly elliptic to oblanceolate, acuminate at the apex, tapering to the winged petiole, sharply serrate, glabrous but sometimes scabrous and sometimes ciliate along the margins, sometimes obscurely 3-veined above the base, to 40 cm long, to 8 cm wide; cauline leaves rapidly becoming smaller above, lanceolate, sessile;

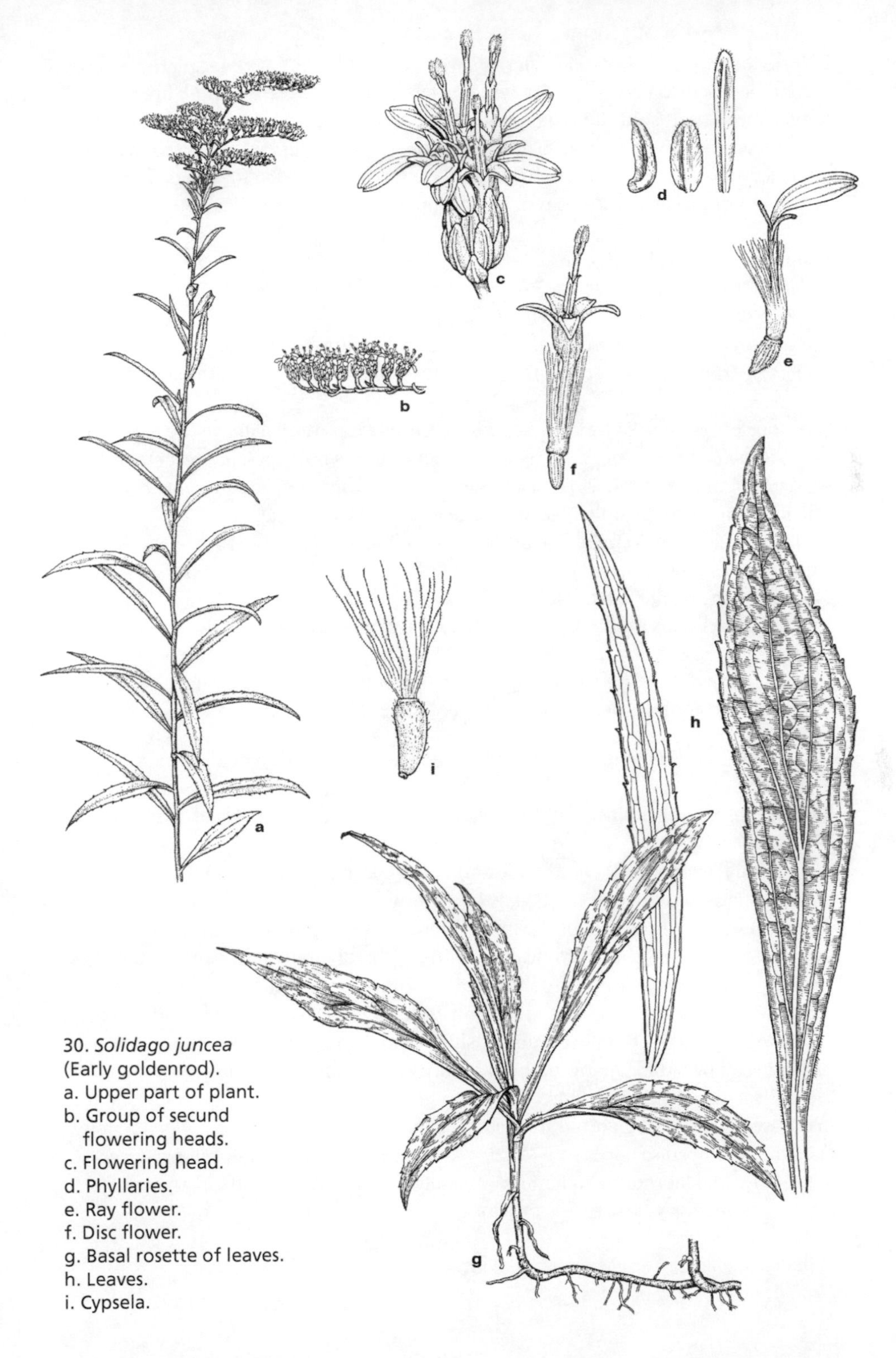

30. *Solidago juncea* (Early goldenrod).
a. Upper part of plant.
b. Group of secund flowering heads.
c. Flowering head.
d. Phyllaries.
e. Ray flower.
f. Disc flower.
g. Basal rosette of leaves.
h. Leaves.
i. Cypsela.

inflorescence a dense pyramidal panicle with recurved branches bearing numerous second heads; involucre 3–5 mm high; phyllaries in 3–5 series, unequal, linear-oblong, acute to obtuse at the apex, appressed, glabrous except for the ciliate margins; receptacle paleate; ray flowers 8–12, (2.5–) 3.0–5.0 mm long, pistillate, yellow; disc flowers 7–10 (–12), 3–4 mm high, bisexual, yellow; cypselae flat, obovoid, finely pubescent or glabrate, 1.2–1.6 mm long; pappus of capillary bristles 2–3 mm long.

Common Name: Early goldenrod.
Habitat: Prairies, glades, dry woods, mesic woods, black oak savannas, roadsides, open disturbed areas.
Range: Nova Scotia to Manitoba, south to Louisiana and Georgia.
Illinois Distribution: Scattered throughout the state; probably in every county.

This species resembles other species whose leaves are 3-veined above the base, such as *S. altissima, S. canadensis,* and *S. gigantea*, but differs by its paleate receptacle, its persistent but often withering basal leaves at flowering time, and its short rhizomes. The leaves are usually only obscurely 3-nerved above the base.

This species also is similar in appearance to *S. glaberrima*, differing by its short rhizomes. *Solidago glaberrima* has much longer rhizomes, and the branches of the inflorescence are glabrous.

A few specimens in Illinois have hirtellous inflorescence branches. These may be called f. *scabrella.*

This is one of the first *Solidago* species to flower in Illinois, beginning as early as mid-June and continuing to flower into September.

20. **Solidago glaberrima** M. Martens, Bull. Acad. Roy. Soc. Bruxelles 8:67. 1841. Fig. 31.
Solidago missouriensis Nutt. var. *fasciculata* Holz. Contr. U. S. Nat. Herb. 1:208. 1892.

Perennial from a thickened caudex and long-creeping rhizomes and slender stolons; stems 1 to several, erect, glabrous throughout except for some puberulence sometimes on the branches of the inflorescence, longitudinally striate; basal leaves early deciduous, less than 1 cm wide; cauline leaves oblanceolate, acute at the apex, tapering to the sessile or short-petiolate base, entire or sometimes serrate, glabrous except for the sometimes ciliate margin, without fascicles of reduced axillary leaves, to 12 cm long, 1.0–2.5 cm wide, obviously 3-veined above the base; inflorescence a rather broad pyramidal panicle, the ascending branches soon recurved, with numerous secund heads; involucre 3–5 mm high; phyllaries oblong-lanceolate, acute to obtuse at the apex, glabrous; receptacle epaleate; ray flowers 7–12, 4–5 mm long, pistillate, yellow; disc flowers up to 20, 3.5–4.5 mm high, bisexual, yellow; cypselae oblongoid, glabrous or sometimes sparsely hispid, 1.0–1.3 mm long; pappus of numerous capillary bristles 2–3 mm long.

Common Name: Eastern Missouri goldenrod.
Habitat: Woods, dry prairies.

Range: Michigan to Washington, south to Texas and Illinois.
Illinois Distribution: Throughout the state.

This species is often considered to be a variety of *S. missouriensis* or even combined with it, but there are a number of subtle differences. The lower leaves of *S. glaberrima* are less than 1 cm wide, and the axils of the leaves do not bear tufts of smaller leaves. The phyllaries of *S. glaberrima* are usually oblong-lanceolate, rather than linear-oblong, and the cypselae are less pubescent or even glabrous and slightly shorter than the cypselae of *S. missouriensis*.

Although I reported *S. missouriensis* previously from Illinois, the specimens on which the report was based are actually *S. juncea*.

The flowers bloom from July to September.

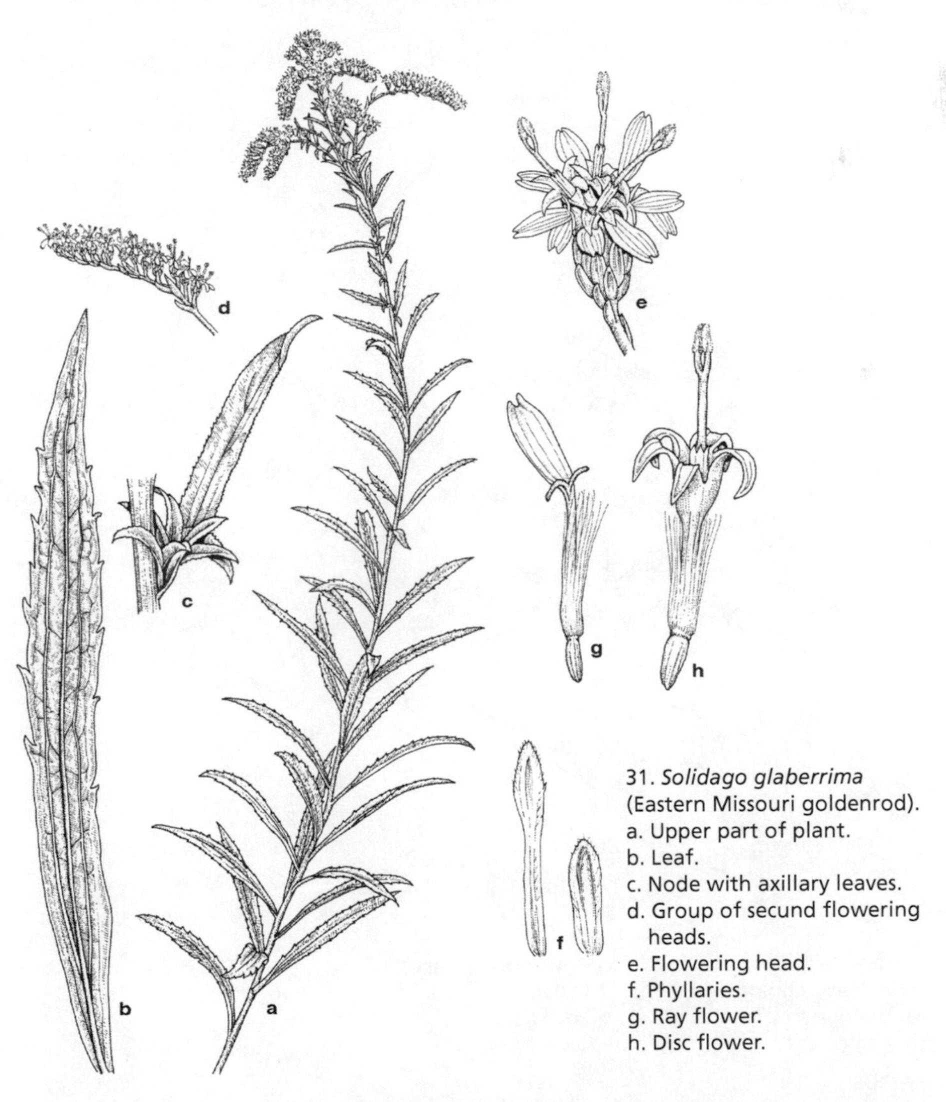

31. *Solidago glaberrima* (Eastern Missouri goldenrod).
a. Upper part of plant.
b. Leaf.
c. Node with axillary leaves.
d. Group of secund flowering heads.
e. Flowering head.
f. Phyllaries.
g. Ray flower.
h. Disc flower.

21. **Solidago ulmifolia** Muhl. ex Willd. Sp. Pl. 3:2060. 1803. Fig. 32.

Perennial from a thickened caudex, without rhizomes; stems 1to a few, erect, to 1.2 (–1.5) m tall, glabrous except on the branches of the inflorescence, with longitudinal striations; basal leaves thin, usually withered by flowering time, ovate to elliptic, acute to acuminate at the apex, tapering to a long petiole, sharply

32. *Solidago ulmifolia* (Elm-leaved goldenrod). a. Upper part of plant. b. Lower leaf. c. Flowering branch. d. Flowering head. e. Phyllary. f. Ray flower. g. Disc flower. h. Cypsela.

serrate, pubescent on both surfaces, sometimes scabrous above, 1-nerved, to 12 cm long, to 5 cm wide; cauline leaves smaller, obovate to elliptic, acute to acuminate at the apex, tapering to the base, short-petiolate or the uppermost sessile, glabrous or sparsely pubescent on the upper surface, hirsute on the lower surface, to 6 cm long, to 2 cm wide; inflorescence an open pyramidal panicle with recurved branches bearing numerous secund heads; involucre 2.5–5.0 mm high; phyllaries in 3 or 4 series, unequal, linear-oblong to narrowly ovate, acute to obtuse at the apex, appressed, glabrous or nearly so, with a diamond-shaped green tip; receptacle epaleate; ray flowers 3–5, 1.5–2.5 mm long, pistillate, yellow; disc flowers 4–7, 2–4 mm high, bisexual, yellow; cypselae flat, obovoid, minutely pubescent, 1.0–1.6 mm long; pappus of capillary bristles 2.0–2.5 mm long.

Common Name: Elm-leaved goldenrod.
Habitat: Dry woods.
Range: Nova Scotia to Minnesota, south to Texas and Florida.
Illinois Distribution: Common throughout the state; in every county.

This species differs from the rare *S. arguta* by not having basal rosettes of leaves at flowering time. It differs from some specimens of *S. gigantea* by having only 1-veined leaves.

Solidago ulmifolia flowers from August to November.

22. **Solidago rugosa** Mill. Gard. Dict., ed. 8:25. 1768. Fig. 33.
Solidago altissima L. var. *rugosa* (Mill.) Torr. Fl. N. Y. 1:363. 1843.
Solidago celtidifolia Small, Fl. S.E.U.S. 1198. 1903.
Solidago rugosa Mill. var. *celtidifolia* (Small) Fern. Rhodora 38:223. 1936.

Perennial from a thickened caudex and long creeping rhizomes; stems 1 to a few, erect, to 2 m tall, densely spreading-villous; lowest leaves withered at flowering time, thin, elliptic to lanceolate, acute to acuminate at the apex, tapering to the sessile or short-petiolate base, with sharp-pointed teeth, glabrous above, pubescent below, 1-veined; cauline leaves thin, lanceolate to elliptic, acute to acuminate at the apex, tapering to the base, sharply serrate or sometimes entire, villous, scarcely rugose, to 12 cm long, to 3 cm wide; inflorescence a pyramidal panicle with recurved branches composed of numerous secund heads; involucre 2.5–5.5 mm high; phyllaries in 3 or 4 series, unequal, linear-oblong, obtuse to acute at the apex, glabrous, with a green midvein, keeled; receptacle epaleate; ray flowers 8–10, 4.5–7.0 mm long, pistillate, yellow; disc flowers 3–7, 2.5–4.0 mm high, bisexual, yellow; cypselae flat, obovoid, 1.0–1.5 mm long, short-pubescent; pappus of capillary bristles 2.0–2.5 mm long.

Common Name: Wrinkled-leaved goldenrod.
Habitat: Bottomland woods, mesic forests, marshes, roadsides.
Range: Manitoba to Washington, south to Texas and Illinois.
Illinois Distribution: Scattered in the southern half of Illinois.

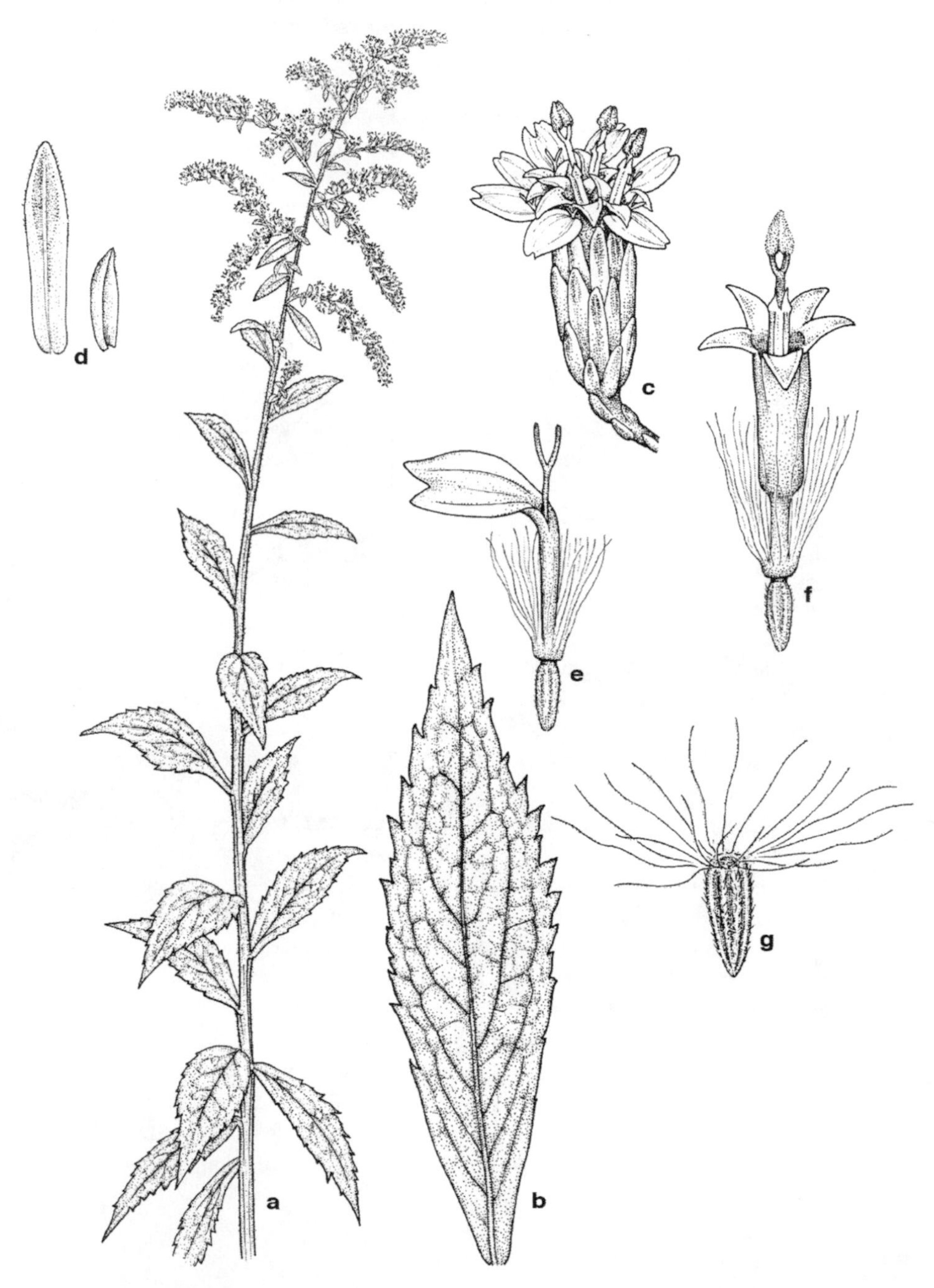

33. *Solidago rugosa* (Wrinkle-leaved goldenrod). a. Upper part of plant. b. Leaf. c. Flowering head. d. Phyllaries. e. Ray flower. f. Disc flower. g. Cypsela.

I am separating this species from *S. aspera*. *Solidago rugosa* has scarcely rugose leaves that are thin and not stiff and with sharp-pointed teeth, and ray flowers 8–10 in number.

The flowers bloom from August to October.

23. **Solidago aspera** Ait. Hort. Kew. 3:212. 1789. Fig. 34.
Solidago rugosa Mill. var. *aspera* (Ait.) Fern. Rhodora 17:7. 1915.
Solidago rugosa Mill. ssp. *aspera* (Ait.) Cronq. Rhodora 49:78. 1947.

Perennial from a thickened caudex and long creeping rhizomes; stems 1 to a few, erect, to 2 m tall, densely hispid; lowest leaves withered at flowering time, thick, firm, elliptic to lanceolate, acute to acuminate at the apex, tapering to the sessile or

34. *Solidago aspera* (Rough wrinkle-leaved goldenrod). a, b. Upper part of plant. c. Stem with leaf. d. Flowering head. e. Ray flower. f. Disc flower. g. Cypsela.

short-petiolate base, serrate with more or less blunt teeth, strongly rugose, scabrous above, pubescent below, to 12 cm long, to 3 cm wide; cauline leaves thin, firm, lanceolate to elliptic, acute to acuminate at the apex, tapering to the base, serrate with blunt teeth, the upper leaves sometimes entire, strongly rugose, short-pubescent, to 12 cm long; involucre 3–6 mm high; phyllaries in 3 or 4 series, unequal, linear to oblong, obtuse to acute at the apex, glabrous, with a green midvein, keeled; receptacle epaleate; ray flowers 6–8, 5–6 mm long, pistillate, yellow; disc flowers 3–7, 2.5–3.5 mm high, bisexual, yellow; cypselae flat, obovoid, 1.0–1.5 mm long, short-pubescent; pappus of capillary bristles 2.0–2.5 mm long.

Common Name: Rough wrinkle-leaved goldenrod.
Habitat: Wet woods, mesic woods, marshes, dry woods, roadsides.
Range: Wisconsin to Washington, south to Texas and Illinois.
Illinois Distribution: Scattered in Illinois but more common in the southern half of the state.

Although most botanists consider this plant to be either a subspecies or variety of *S. rugosa*, I am maintaining it as a separate species because of its hispid stems and its thick, firm, scabrous, and obviously rugose leaves with blunt-tipped teeth. It also has fewer rays per flowering head than *S. rugosa*.

The flowers bloom in August and September.

24. **Solidago drummondii** Torr. & Gray, Fl. N. Am. 2:217. 1842. Fig. 35.

Perennial from a stout caudex and usually stout rhizomes; stems 1-several, erect, to 1 m tall, with spreading soft pubescence, longitudinally striate; basal leaves absent at flowering time, broadly ovate to broadly elliptic, acute to acuminate at the apex, rounded at the short-petiolate base, serrate, 3-veined, densely and softly pubescent on both surfaces, to 10 cm long, to 7 cm wide; middle and upper leaves smaller and usually entire or nearly so; inflorescence an open pyramidal panicle, with numerous secund heads on recurving branches; involucre 3.0–4.5 mm high; phyllaries in 3 or 4 series, unequal, oblong, obtuse at the apex, appressed, glabrous except for the ciliate margins; receptacle epaleate; ray flowers 4–7, 2–5 mm long, pistillate, yellow; disc flowers 5–9, 4–5 mm high, bisexual, yellow; cypselae flat, obovoid, with short pubescence, 1.5–2.5 mm long; pappus of capillary bristles 2.0–2.5 mm long.

Common Name: Ozark goldenrod.
Habitat: Crevices of limestone bluffs.
Range: Arkansas, Illinois, Louisiana, Missouri.
Illinois Distribution: Mostly in counties along the Mississippi River from Pike County southward; also Hardin and Pulaski counties.

This species is readily distinguished from other species of *Solidago* by its broadly ovate basal leaves and its densely, softly spreading-pubescent stems and leaves.

Solidago drummondii usually occurs in crevices of limestone cliffs along the Mississippi River. It flowers from August to November.

35. *Solidago drummondii* (Ozark goldenrod).

a. Upper part of plant.
b. Flowering head.
c. Phyllary.
d. Ray flower.
e. Disc flower.
f. Cypsela.

25. **Solidago canadensis** L. Sp. Pl. 2:878. 1753. Fig. 36.
Solidago canadensis L. var. *hargeri* Fern. Rhodora 17:11–12. 1915.

Perennial from long creeping or sometimes rather short rhizomes; stems 1 to several, erect, to 1.5 (–2.0) m tall, densely short-hairy to pilose except for the sometimes glabrous base, longitudinally striate; basal leaves absent at flowering time; cauline leaves crowded, lanceolate to oblanceolate, acute to acuminate at the apex, tapering to the sessile or very short-petiolate base, to 15 cm long, to 2.5 cm wide,

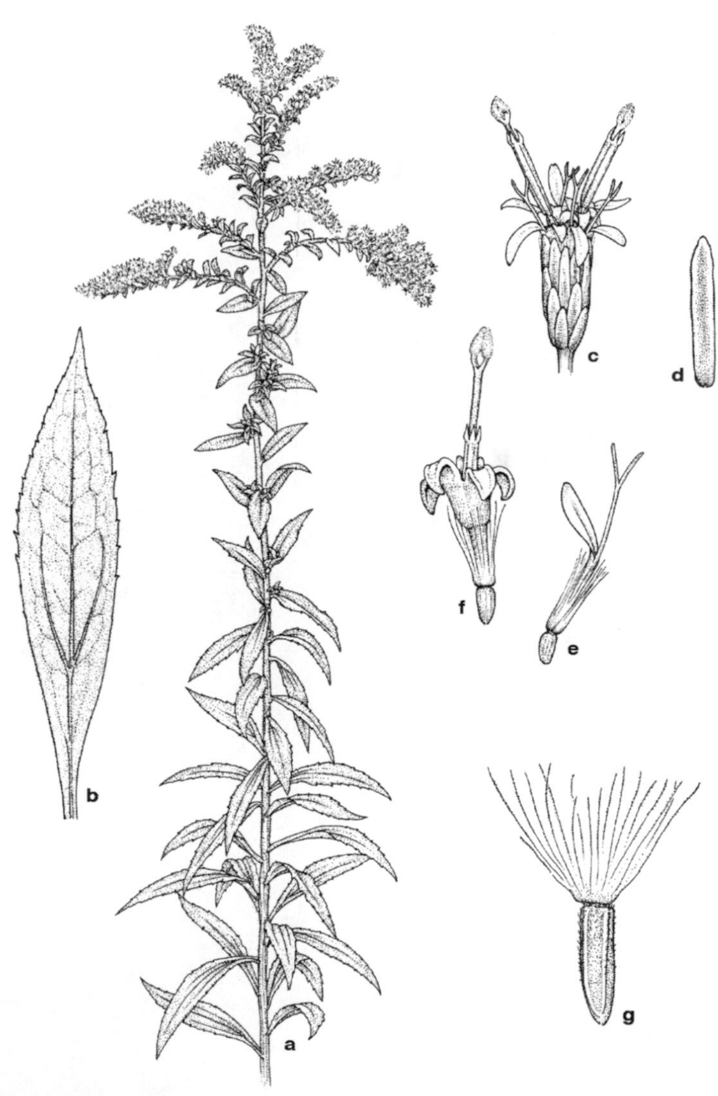

36. *Solidago canadensis* (Canada goldenrod). a. Upper part of plant. b. Leaf. c. Flowering head. d. Phyllary. e. Ray flower. f. Disc flower. g. Cypsela.

the upper leaves sessile and progressively smaller and often nearly entire, usually sharply serrate, glabrous or with swollen-based hairs on the upper surface, densely pubescent on the lower surface, or sometimes merely pilose on the veins, 3-veined above the base; inflorescence a dense pyramidal panicle with numerous slender heads, those on the recurved lower branches often secund; involucre 2–3 mm high; phyllaries in 3–5 series, unequal, linear to lanceolate, the margins sparsely pilose, otherwise glabrous, keeled, yellowish, mostly all long-attenuate; receptacle epaleate; ray flowers 6–12, 2–3 mm long, pistillate, yellow; disc flowers 2–5, 2.3–2.7 mm long, bisexual, yellow; cypselae hirtellous, 1.0–1.5 mm long, obovoid; pappus of capillary bristles, 1.5–2.0 mm long.

Common Name: Canada goldenrod; old field goldenrod.
Habitat: Prairies, dry woods, savannas, old fields, open disturbed areas, along rivers and streams, roadsides, pastures.
Range: Newfoundland to Manitoba, south to Nebraska, Illinois, Kentucky, and Virginia.
Illinois Distribution: Throughout the state; in every county.

Two varieties occur in Illinois. Typical var. *canadensis* has stems densely pubescent throughout. Var. *hargeri* has stems that are moderately to densely pubescent but usually glabrous near the base.

Solidago canadensis differs from *S. altissima* by its shorter involucre, shorter ray flowers, shorter disc flowers, and shorter pappus bristles. Some botanists do not distinguish these two closely related species as separate species.

This species blooms from late July through October.

26. **Solidago altissima** L. Sp. Pl. 2:878–879. 1753. Fig. 37.
Solidago procera Ait. Hort. Kew. 3:211. 1789.
Solidago canadensis L. var. *procera* (Ait.) Torr. & Gray, Fl. N. Am. 2:221. 1841.
Solidago canadensis L. var. *gilvocanescens* Rydb. Contr. U.S. Natl. Herb. 3:162. 1895.
Solidago gilvocanescens (Rydb.) Smyth, Trans. Kans. Acad. Sci. 16:161. 1899.
Solidago altissima L. var. *procera* (Ait.) Fern. Rhodora 10:92. 1908.
Solidago canadensis L. ssp. *gilvocanescens* (Rydb.) A. Love & D. Love, Taxon 31:358. 1982.
Solidago altissima L. var. *gilvocanescens* (Rydb.) Semple, Phytologia 58:430. 1985.
Solidago altissima L. ssp. *gilvocanescens* (Rydb.) Semple, Sida 20:1606. 2003.

Perennial from long creeping or sometimes rather short rhizomes; stems 1 to several, erect, to 2.0 (–2.5) m tall, densely short-hairy with curved hairs, often at length nearly glabrous at the base, longitudinally striate; basal leaves absent at flowering time; cauline leaves crowded, lanceolate to oblanceolate, acute to acuminate at the apex, tapering to the sessile or very short-petiolate base, to 15 cm long, to 2.5 cm wide, the upper leaves progressively smaller and often nearly entire, usually sharply serrate but sometimes obscurely toothed, pubescent with curved or spreading, swollen-based hairs, or with short, curved hairs, sometimes scabrous on the upper

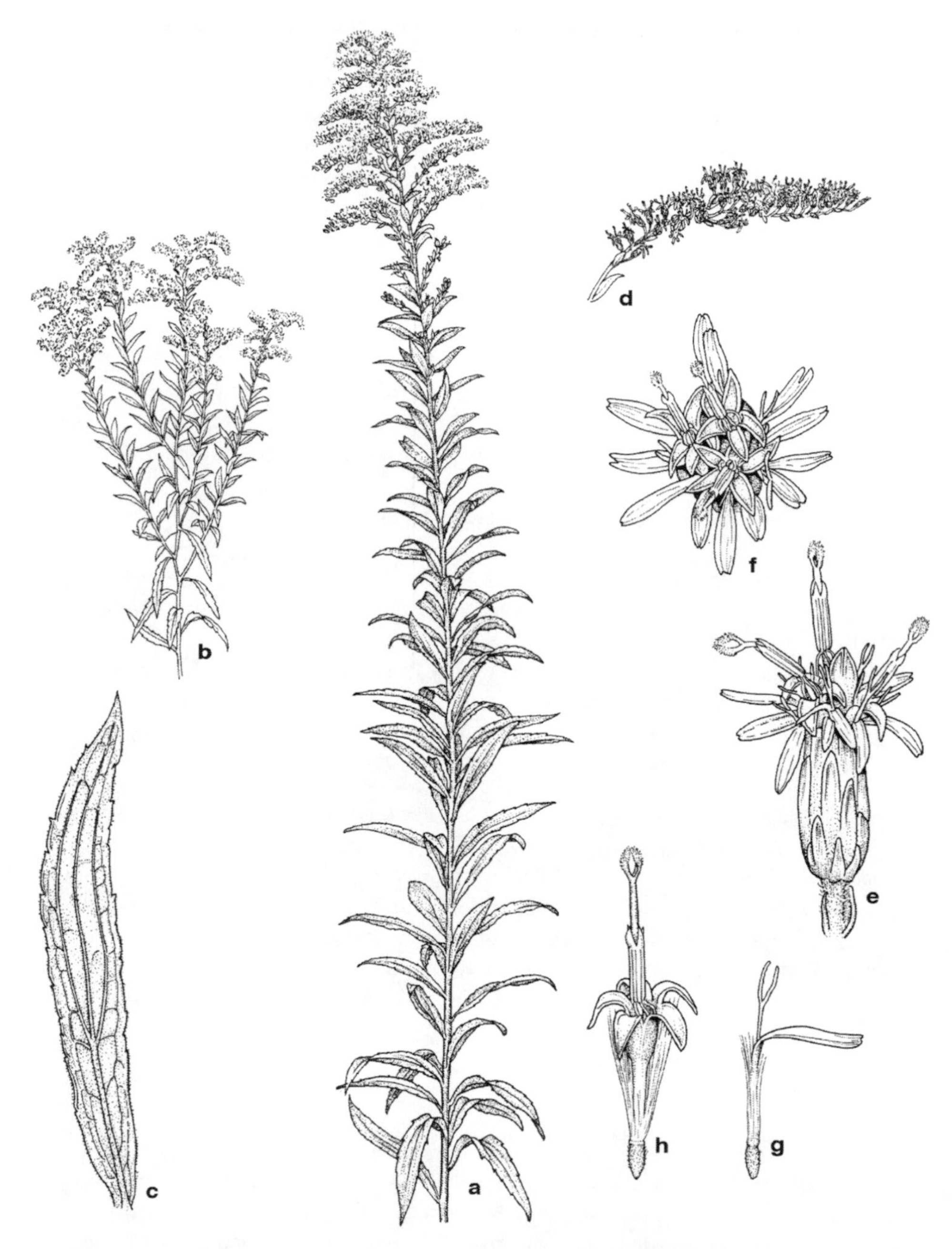

37. *Solidago altissima* (Tall goldenrod).
a. Habit.
b. Upper part of plant.
c. Leaf.
d. Branch with flowering heads.
e. Flowering head, side view.
f. Flowering head, face view.
g. Ray flower.
h. Disc flower.

surface and along the margins, densely short-hairy on the lower surface, rarely hairy on the midvein below, 3-veined above the base; inflorescence a dense pyramidal terminal panicle with numerous slender heads, those on the recurved lower branches often secund; involucre 3.0–4.5 mm high; phyllaries in 3–6 series, unequal, lanceolate, glabrous or hirtellous, keeled, yellowish, the outer ones acute, the inner ones obtuse to subacute; receptacle epaleate; ray flowers 10–16, (1–) 2–3 (–4) mm long, pistillate, yellow; disc flowers 3–7, 3.0–3.5 mm long, bisexual, yellow; cypselae hirtellous, 0.8–1.4 mm long, obovoid; pappus of capillary bristles, 2.8–3.4 mm long.

Common Name: Tall goldenrod.
Habitat: Mesic woods, dry woods, prairies, old fields, pastures, fens, roadsides.
Range: Nova Scotia to Saskatchewan, south to California, Texas, and Florida.
Illinois Distribution: Throughout the state; in every county.

Typical variety *altissima* has the upper leaf surface rough-pubescent with bulbous-based hairs, or sometimes hairy only on the midvein beneath. Variety *gilvocanescens*, the more common variety in Illinois, has leaves with short, curved hairs that are not bulbous-based. As a result, the leaves of var. *gilvocanescens* are scarcely if at all scabrous.

Although *S. altissima* and *S. canadensis* are considered to be separate species in this work, there are some specimens that intergrade between the two so that the distinctness of these specimens is not obvious. Usually, *S. altissima* has an involucre that is 3.0–4.5 mm high, has 10–16 ray flowers that are 3–4 mm long, and has 3–7 disc flowers that are 3.0–3.5 mm high, while *S. canadensis* usually has an involucre that is 2–3 mm high, 6–12 ray flowers that are 2–3 mm long, and 2–5 disc flowers that are 2.3–2.7 mm high.

Solidago altissima flowers from July to October.

27. **Solidago gigantea** Ait. Hort. Kew. 3:211. 1789. Fig. 38.
Solidago gigantea Ait. var. *leiophylla* Fern. Rhodora 41:457. 1939.

Perennial from long creeping rhizomes; stems 1 to several, erect, to 2 m tall, glabrous except in the puberulent inflorescence, often glaucous, longitudinally striate; basal leaves absent; lower cauline leaves lanceolate to elliptic, acute to acuminate at the apex, tapering to the sessile or rarely short-petiolate base, sharply serrate, glabrous on both surfaces or sometimes pilose on the veins beneath, the margins usually somewhat scabrous, 3-nerved above the base, to 15 cm long, to 3.5 (–4.0) cm wide, the upper leaves smaller and sessile; inflorescence a dense terminal pyramidal panicle, the lower branchlets recurved with the heads secund, the branchlets of the panicle puberulent; involucre 2.5–4.0 mm high; phyllaries in 3–5 series, unequal, linear to lanceolate, obtuse to acute, appressed, glabrous except for the sparsely pubescent margins, more or less green-tipped; receptacle epaleate; ray flowers (8–)10–17, 2–5 mm long, pistillate, yellow; disc flowers 3–8 (–10), 3–4 mm long, bisexual, yellow; cypselae obovoid, glabrous or sparsely pubescent, 1.3–1.6 mm long; pappus of numerous capillary bristles 2.0–2.5 mm long.

Common Name: Late goldenrod; tall goldenrod.
Habitat: Wet woods, mesic woods, around ponds and lakes, fens, bogs, floodplains, moist prairies, uncommon in drier habitats.
Range: Nova Scotia to Saskatchewan, south to Colorado, Texas, and Florida.
Illinois Distribution: Throughout the state; in every county.

Many of our plants have completely glabrous leaves and have been called var. *leiophylla*.

This species has the general growth form of *S. altissima* and *S. canadensis*, and all three have leaves that are 3-veined above the base. *Solidago gigantea* differs readily by its stems that are glabrous below the inflorescence. The stems are often glaucous.

This species is almost always associated with moist to wet habitats, while *S. altissima* and *S. canadensis*, although primarily upland species, may also be found in similar wet habitats.

The flowers bloom from August to October.

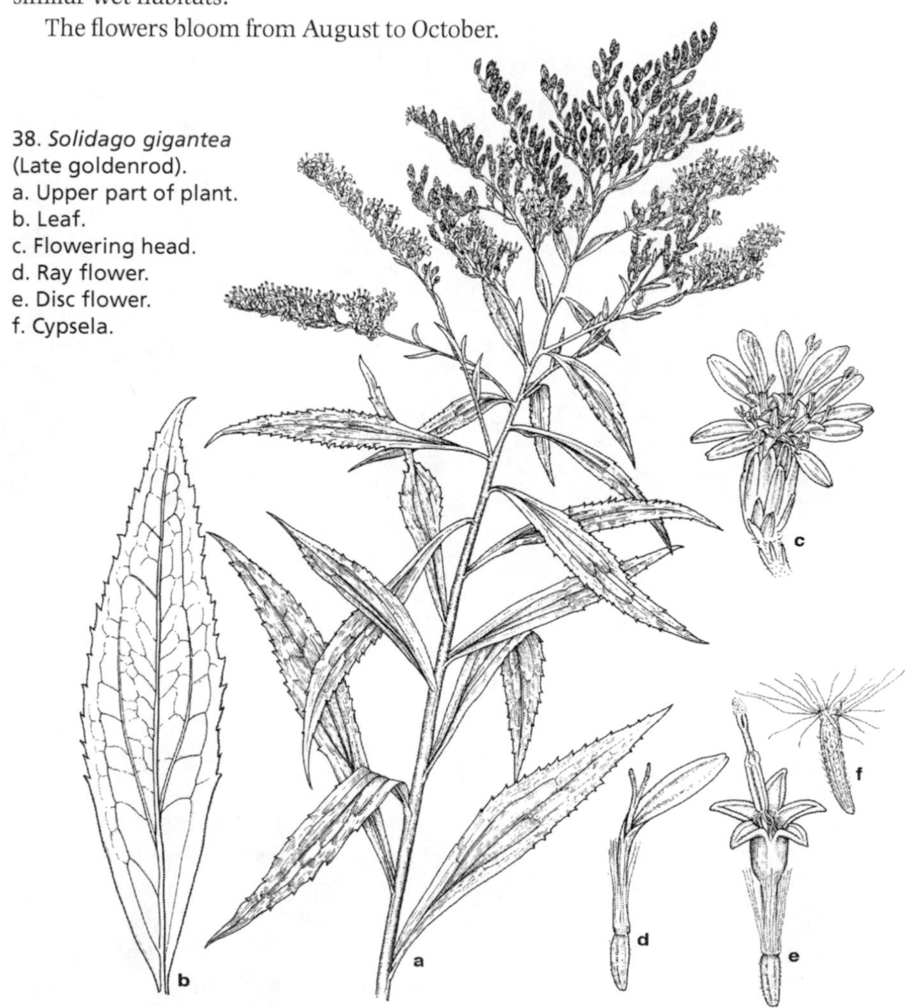

38. *Solidago gigantea* (Late goldenrod).
a. Upper part of plant.
b. Leaf.
c. Flowering head.
d. Ray flower.
e. Disc flower.
f. Cypsela.

28. **Solidago nemoralis** Ait. Hort. Kew. 3:213. 1789. Fig. 39.

Perennial from a stout caudex and short rhizomes; stems 1 to a few, erect, to 1 m tall, finely gray-puberulent throughout, with longitudinal striations; basal leaves persistent at flowering time, spatulate to broadly oblanceolate, acute at the apex, tapering to the subsessile or short-petiolate base, crenate to serrulate, puberulent on both surfaces, to 25 cm long, to 4 cm wide, 1-veined; middle and upper leaves

39. *Solidago nemoralis* (Gray goldenrod).

a. Lower part of plant.
b. Upper part of plant.
c. Flowering head.
d. Phyllary.
e. Ray flower.
f. Disc flower.

gradually decreasing in size, narrowly oblanceolate, acute at the apex, tapering to the sessile base, often entire or nearly so, to 8 cm long, to 3 cm wide, 1-nerved; inflorescence a narrow or sometimes open panicle, with recurved branches bearing numerous secund heads, the branches often viscidulous; involucre 3.0–4.5 mm high; phyllaries in 3 or 4 series, unequal, linear, obtuse, glabrous except for the ciliate margins; receptacle epaleate; ray flowers 5–9, 3.0–4.5 mm long, pistillate, yellow; disc flowers 3–9, 2.5–4.5 mm high, bisexual, yellow; cypselae flat, obovoid, strigose, 2.0–2.5 mm long; pappus of capillary bristles 2.5–3.0 mm long.

Common Name: Gray goldenrod.
Habitat: Old fields, pastures, dry prairies, dry woods, dunes around Lake Michigan.
Range: New Brunswick to North Dakota, south to New Mexico, Texas, and Illinois.
Illinois Distribution: Common throughout the state; probably in every county.

This species, whose leaves and stems are slightly gray because of the pubescence, has 1-veined leaves that range from being crenate to serrulate. Its shorter involucres, wider leaves, and strigose cypselae distinguish it from *S. decemflora*, although the latter is often considered to be a variety or subspecies of *S. nemoralis*.

The flowers bloom from July to early November.

29. **Solidago decemflora** DC. Prodr. 5:332. 1836. Fig. 40.
Solidago longipetiolata Mack. & Bush, Trans. Acad. Sci. St. Louis 12:87–88. 1902.
Solidago nemoralis DC. var. *longipetiolata* (Mack. & Bush) Palmer & Steyerm. Ann. Mo. Bot. Gard. 22:660. 1935.
Solidago nemoralis DC. ssp. *longipetiolata* (Mack. & Bush) Brammall ex Semple, Phytologia 58:430. 1985.

Perennial from a stout caudex and short rhizomes; stems 1 to a few, erect, to 1 m tall, finely puberulent throughout, with longitudinal striations; basal leaves persistent at flowering time, linear-lanceolate to narrowly lanceolate, acute at the apex, tapering to the base, entire to occasionally sharply toothed near the tip, puberulent on both surfaces, to 15 cm long, to 2.5 cm wide, 1-veined; middle and upper leaves decreasing in size, linear to linear-lanceolate, acute at the apex, tapering to the sessile base, to 6 cm long, to 2 cm wide, 1-veined; inflorescence a narrow to sometimes slightly open panicle, with recurved branches bearing numerous secund heads; involucre 4.5–6.0 (–6.5) mm high; phyllaries in 3 or 4 series, unequal, linear-oblong, obtuse to acute at the apex, glabrous except for the ciliate margins; receptacle epaleate; ray flowers 5–9, 3–4 mm long, pistillate, yellow; disc flowers 3–9, 2–3 mm high, bisexual, yellow; cypselae flat, obovoid, sericeous, 1.8–2.2 mm long; pappus of capillary bristles 2.5–3.0 mm long.

Common Name: Gray goldenrod.
Habitat: Old fields.
Range: Ontario to British Columbia, south to New Mexico, Missouri, and Illinois.
Illinois Distribution: Scattered in the northern fourth of Illinois.

Although sometimes considered a variety or subspecies of *S. nemoralis*, the narrower leaves, sericeous cypselae, taller involucres, and slightly smaller cypselae distinguish it. I choose to recognize *S. decemflora* as a distinct species since it usually is easily differentiated in the field.

The flowers bloom from August to late October.

40. *Solidago decemflora* (Gray goldenrod).
a. Upper part of plant.
b. Middle stem with leaves.
c. Flowering head, side view.
d. Flowering head, face view.
e. Ray flower.
f. Disc flower.
g. Cypsela.

30. **Solidago radula** Nutt. Journ. Acad. Nat. Sci. Phila. 7:102. 1834. Fig. 41.
Solidago laeta Greene, Pittonia 5:138–139. 1903.
Solidago radula Nutt. var. *laeta* (Greene) Fern. Rhodora 38:228. 1936.
Solidago radula Nutt. var. *stenolepis* Fern. Rhodora 38:228–229. 1936.

Perennial from a thickened caudex and sometimes from long creeping rhizomes; stems 1 to several, erect, to 1 m tall, hirtellous to hirsute, with longitudinal striations; leaves rigid, the basal withered by flowering time, acute at the apex, tapering to a winged petiole, elliptic-lanceolate to elliptic, finely serrate to entire, usually pubescent on both surfaces, 3-nerved, to 10 cm long, to 2.5 cm wide; middle and upper leaves lanceolate to elliptic, acute to subacute at the apex, tapering to the sessile or short-petiolate base, entire to crenate, usually 3-veined, usually hirsutulous on both surfaces, to 8 cm long, to 3 cm wide; inflorescence a dense pyramidal panicle, the branches stiff, often ascending, pubescent, the heads secund; involucre 3.0–5.5 mm high; phyllaries in 3 or 4 series, unequal, firm, linear-lanceolate to oblong-ovate, acute to obtuse at the apex, appressed, glabrous with ciliate margins, with a green, diamond-shaped tip; receptacle epaleate; ray flowers 4–8, 2–4 mm

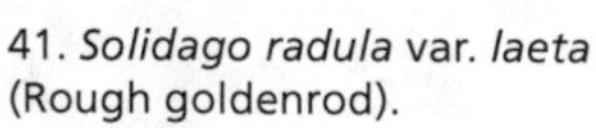

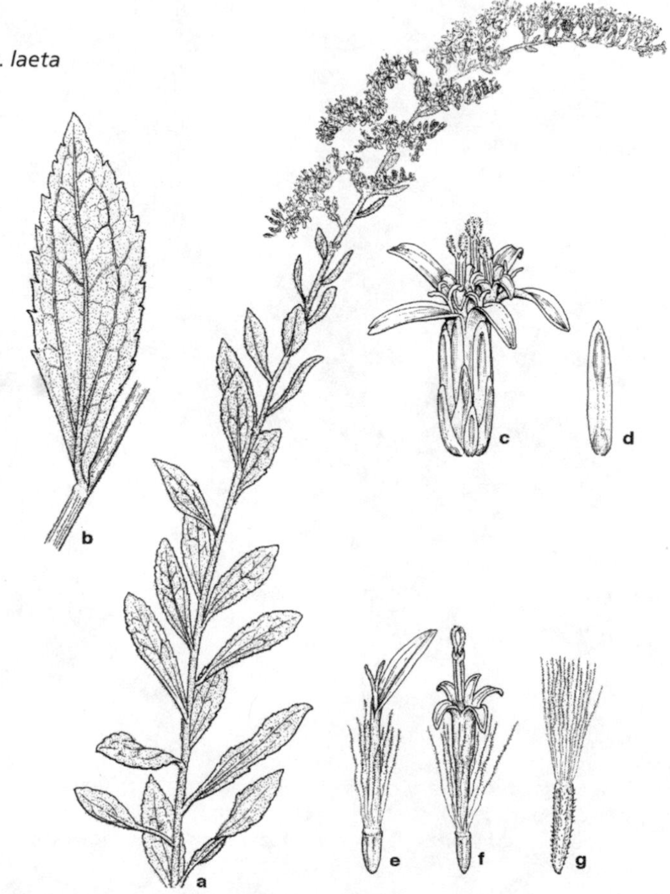

41. *Solidago radula* var. *laeta* (Rough goldenrod).
a. Upper part of plant.
b. Leaf.
c. Flowering head, side view.
d. Phyllary.
e. Ray flower.
f. Disc flower.
g. Cypsela.

long, pistillate, yellow; disc flowers 4–8 (–10), 2.5–4.0 mm high, bisexual, yellow; cypselae flat, obovoid, 1.5–2.5 mm long, short-hairy; pappus of capillary bristles 2.5–3.0 mm long.

Common Name: Western rough goldenrod.
Habitat: Mesic woods, dry woods, savannas, prairies, roadsides.
Range: Kentucky to Kansas, south to Texas and Louisiana; also Georgia, North Carolina, and South Carolina.
Illinois Distribution: Confined to the western half of the state, extending north to Henderson County.

This species is distinguished by its hirsute to hirtellous stems and leaves and its prominently 3-veined leaves. It differs from the similar *S. drummondii* by its hirsute, petiolate leaves. Specimens with broader and shorter phyllaries have been called var. *laeta*, while specimens with very narrow phyllaries have been called var. *stenolepis*.

Solidago radula flowers from August to October.

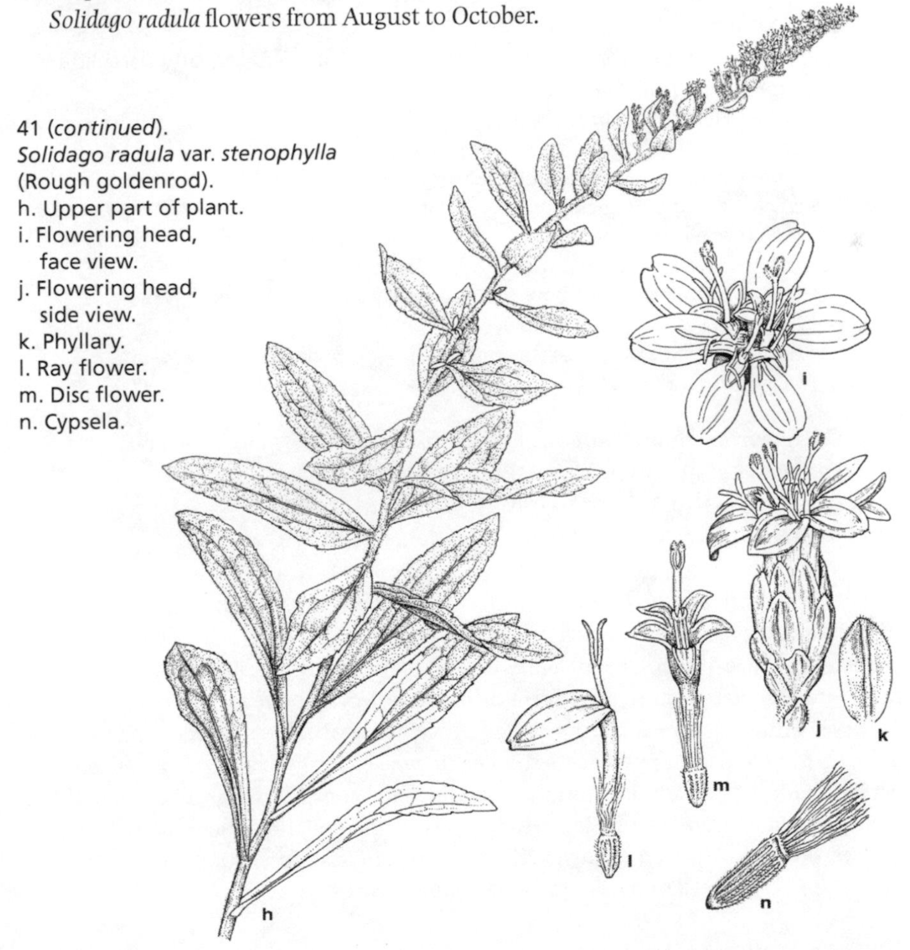

41 (*continued*).
Solidago radula var. *stenophylla* (Rough goldenrod).
h. Upper part of plant.
i. Flowering head, face view.
j. Flowering head, side view.
k. Phyllary.
l. Ray flower.
m. Disc flower.
n. Cypsela.

9. **Oligoneuron** Small

Perennial herbs with short or long rhizomes; stems 1 to several, erect to ascending, finely longitudinally grooved, glabrous or pubescent; basal leaves usually persistent at flowering time and often with a rosette of leaves present as well, usually on long petioles, glabrous or pubescent, entire or serrate; cauline leaves progressively smaller toward the top of the stem, short-petiolate or sessile, glabrous or pubescent; inflorescence a round- or flat-topped corymb, the branches often short and compact, the heads solitary or in small clusters, radiate; involucre campanulate; phyllaries in 3–6 series, unequal, appressed, 1- to 7-nerved, linear to oblong to ovate, glabrous or pubescent; receptacle epaleate; ray flowers pistillate, yellow or cream or white; disc flowers perfect, yellow or white; cypselae obovoid, usually not flat, glabrous or pubescent; pappus of numerous capillary bristles in 2 series, the outer shorter and not thickened at the tip, the inner longer and thickened at the tip.

There are six species in this genus, all in North America.

I am accepting *Oligoneuron* as a distinct genus on the basis of its round- or flat-topped corymbs and its pappus bristles in 2 unequal series. The ray flowers may be yellow or cream or white.

Some botanists who have split out several genera from *Solidago* and *Aster* still maintain the species of *Oligoneuron* in *Solidago*, a view not followed here.

1. Leaves ovate to elliptic, scabrous .... 4. *O. rigidum*
1. Leaves linear to lanceolate, glabrous except sometimes near the apex.
  2. Inflorescence branches puberulent; leaves longitudinally folded....... 3. *O. riddellii*
  2. Inflorescence branches glabrous; leaves flat.
    3. Ray flowers white.. 1. *O. album*
    3. Ray flowers yellow or cream.
      4. Leaves usually serrate above the middle; rays up to 5 mm long, yellow .... 2. *O. ohioense*
      4. Leaves entire or nearly so throughout; rays 5–8 mm long, cream............. .... 5. *O.* X *lutescens*

1. **Oligoneuron album** (Nutt.) G. L. Nesom, Phytologia 75:28. 1993. Fig. 42.
Inula ptarmicoides Nutt. Gen. N. Am. Pl. 2:152. 1818.
Aster ptarmicoides (Nutt.) Torr. & Gray, Fl. N. Am. 2:160. 1842.
Solidago ptarmicoides (Nutt.) B. Boivin, Phytologia 23:21. 1972.

Perennial from a branched, thickened caudex; stems erect, slender, stiff, to 60 cm tall, glabrous or sometimes hirtellous in the upper half; previous year's leaves often persistent at base of plant; lower leaves of the current season coriaceous, firm, linear to linear-lanceolate, obtuse to acute at the apex, tapering to the sessile base, entire or very sparingly serrate, glabrous but often scabrous, usually 3-nerved, up to 20 cm long, up to 1 cm wide; upper leaves smaller, linear, sessile; inflorescence a flat-topped corymb with 3–75 heads on strongly ascending branches; involucre 4–7 mm high, nearly as wide, turbinate; phyllaries in 3–5 series, unequal, the inner oblong to lanceolate, firm, acute to obtuse, 1-nerved, greenish, glabrous; receptacle epaleate; ray flowers 4–7 (–12), 5–8 mm long, pistillate, white; disc flowers 8–15 (–30), 3.5–5.0 mm long, bisexual, white; cypselae obconic, flat, glabrous, several-nerved, 1.0–1.5 mm long; pappus of numerous capillary bristles enlarged at the tip, 3.5–4.0 mm long.

Common Name: Stiff aster.
Habitat: Prairies, sandy soil, dry calcareous prairies.
Range: Quebec to Saskatchewan, south to Colorado, Oklahoma, Missouri, Illinois, and Ohio; South Carolina.
Illinois Distribution: Occasional in the northern half of Illinois.

In the past, this species has been placed in *Aster* or in *Solidago*. Because of its flat-topped inflorescence and reticulate-veined leaves, I am placing it in *Oligoneuron*. It differs from other members of *Oligoneuron* by its white rays.

The stiff stems and coriaceous leaves also distinguish this species.

Because Nuttall's *Inula alba* has priority over *Aster ptarmicoides*, the correct epithet for this species is *album*.

The flowers appear from June to October.

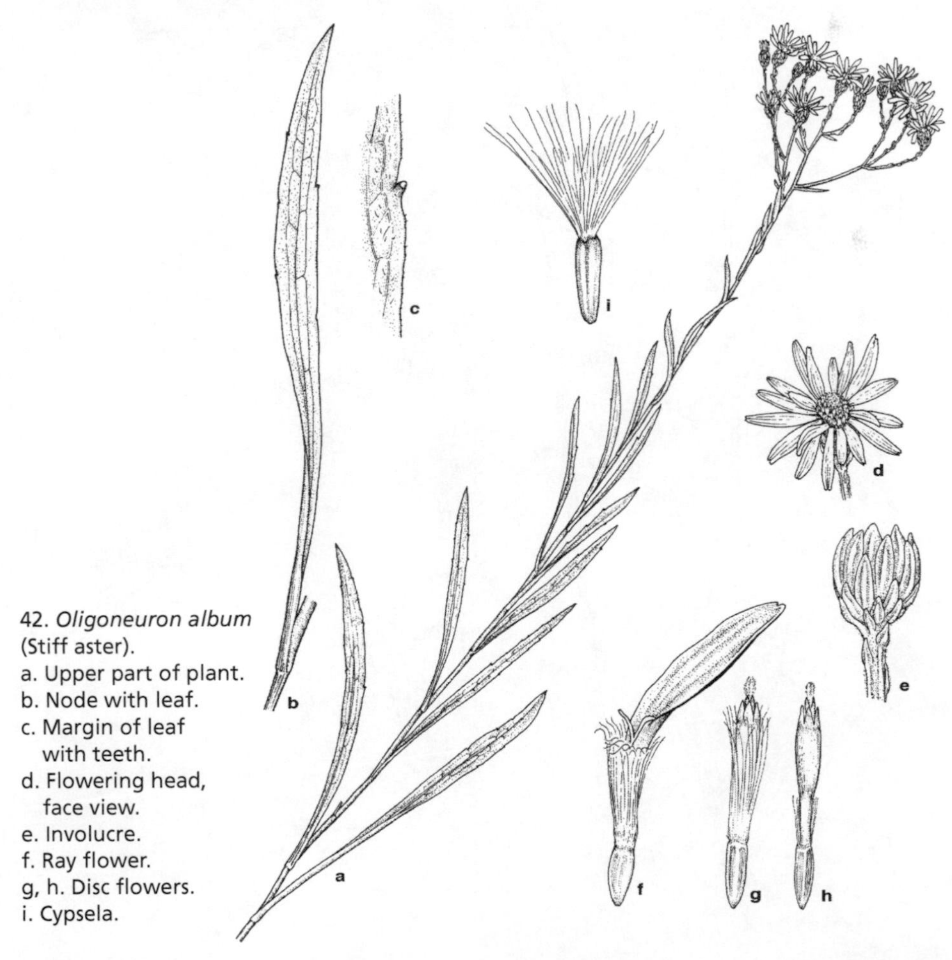

42. *Oligoneuron album* (Stiff aster).
a. Upper part of plant.
b. Node with leaf.
c. Margin of leaf with teeth.
d. Flowering head, face view.
e. Involucre.
f. Ray flower.
g, h. Disc flowers.
i. Cypsela.

2. **Oligoneuron ohioense** (Riddell) G. N. Jones, Trans. Ill. State Acad. Sci. 35:63. 1942. Fig. 43.
Solidago ohioensis Riddell, W. J. Med. Phys. Sci. 8:499. 1835.

Perennial from a thickened caudex; stems erect, to 90 cm tall, glabrous or nearly so; basal leaves persistent, lance-elliptic to narrowly ovate, subacute to acute at the apex, tapering to the base to a long petiole, entire or sparsely serrate, glabrous except for the scabrous and sometimes ciliate margins, up to 20 cm long, up to 5 cm wide; middle and upper cauline leaves progressively smaller, lance-elliptic to elliptic, flat, 1-nerved, up to 6 cm long, up to 1.5 cm wide; inflorescence a terminal,

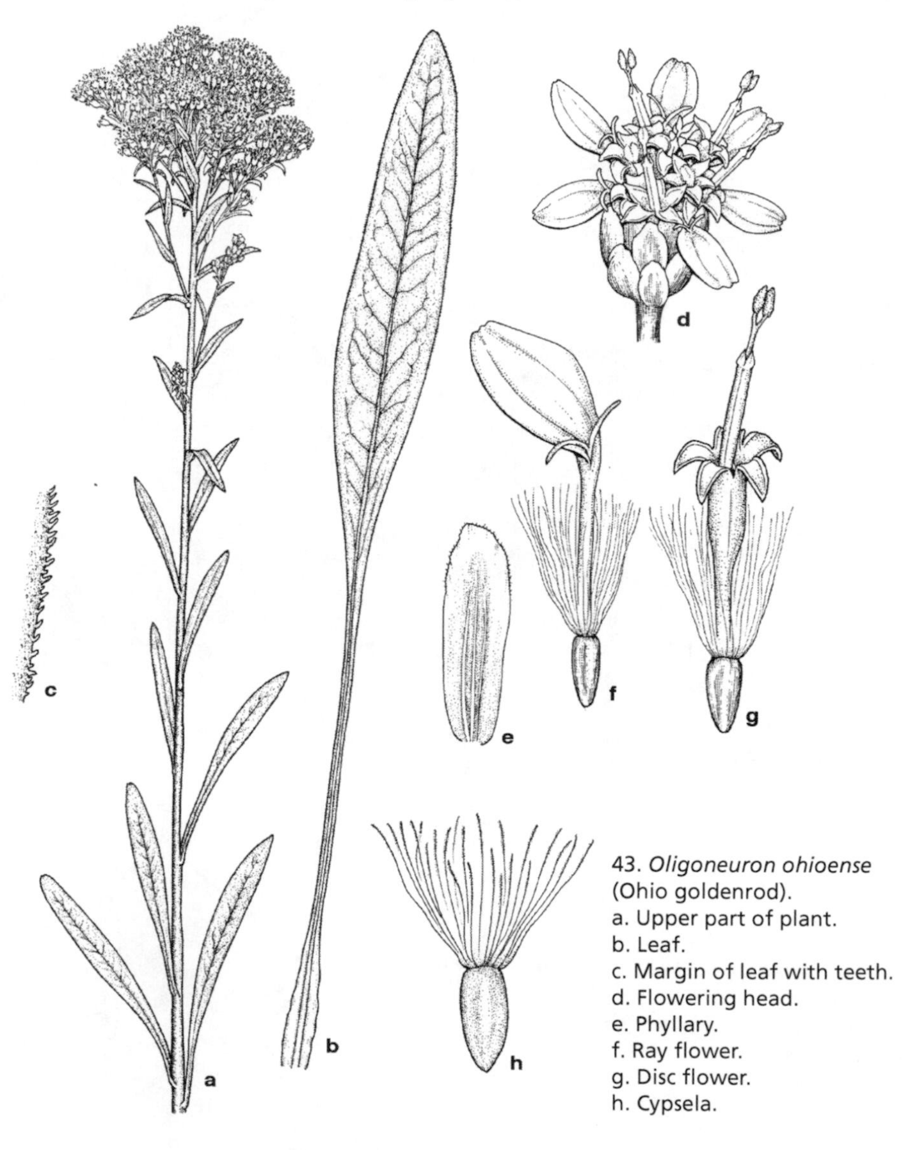

43. *Oligoneuron ohioense* (Ohio goldenrod).
a. Upper part of plant.
b. Leaf.
c. Margin of leaf with teeth.
d. Flowering head.
e. Phyllary.
f. Ray flower.
g. Disc flower.
h. Cypsela.

flat-topped corymb with 50–500 heads on glabrous peduncles; involucre 4–6 mm high; phyllaries in 3 or 4 series, unequal, firm, elliptic, obtuse, glabrous, striate; receptacle epaleate; ray flowers 6–8, 4–5 mm long, pistillate, yellow; disc flowers up to 20, 4.0–4.5 mm long, bisexual, yellow; cypselae obconic, glabrous, 3- or 5-angled, obscurely nerved, 1.3–2.2 mm long; pappus of numerous white capillary bristles enlarged at the tip, 2.5–3.0 mm long.

Common Name: Ohio goldenrod.
Habitat: Calcareous fens, low sand flats, other moist soil habitats.
Range: New York to Ontario, south to Illinois and Indiana.
Illinois Distribution: Occasional in northeastern Illinois, extending south to Peoria and Woodford counties.

This species is often placed in *Solidago*, but I prefer to include it in *Oligoneuron*.

It resembles *O. riddellii*, differing by its flat leaves and glabrous branches of the inflorescence.

Oligoneuron ohioense flowers from June to October.

3. **Oligoneuron riddellii** (Frank) Rydb. Fl. Plains N. Am. 799. 1932. Fig. 44.
Solidago riddellii Frank, W. J. Med. Phys. Sci. 8:499. 1835.

Perennial from a thickened caudex and sometimes from long, creeping rhizomes; stems erect, stout, to 1 m tall, glabrous except rough-hairy in the inflorescence; basal leaves absent at flowering time, narrowly linear-lanceolate, acute and recurved at the apex, folded lengthwise, tapering to the base to a long winged petiole, entire, glabrous except for the scabrous margins, up to 50 cm long, up to 2.5 cm wide; middle and upper leaves crowded, linear-lanceolate, obtuse to acute and recurved at the apex, tapering to the sessile or sometimes clasping base, usually folded lengthwise, mostly 3-nerved, up to 6 cm long, up to 1 cm wide; inflorescence a terminal, flat-topped or round-topped corymb with 50–500 crowded heads on rough-hairy peduncles; involucre 4.5–6.0 mm high; phyllaries in 3 or 4 series, unequal, firm, elliptic, obtuse, glabrous, striate; receptacle epaleate; ray flowers 7–9, 4.0–5.5 mm long, pistillate, yellow; disc flowers up to 10, 4.5–6.0 mm long, bisexual, yellow; cypselae obconic, glabrous or nearly so, 5- or 7-nerved, 1.3–2.0 mm long; pappus of numerous white capillary bristles enlarged at the tip, 3.5–4.0 mm long.

Common Name: Riddell's goldenrod.
Habitat: Prairies, calcareous fens, other moist soil habitats.
Range: Ontario to Manitoba, south to Missouri, Illinois, Indiana, and Ohio; New York.
Illinois Distribution: Occasional in the northern half of Illinois, rare in the southern half.

Like the preceding species, this plant is often placed in the genus *Solidago*. It differs from the similar-appearing *O. ohioense* by its leaves that are folded lengthwise and by the pubescence on the branches of the inflorescence.

This species flowers from August to November.

44. *Oligoneuron riddellii* (Riddell's goldenrod).
a. Upper part of plant.
b. Flowering head.
c. Phyllary.
d. Ray flower.
e. Disc flower.
f. Cypsela.

4. **Oligoneuron rigidum** (L.) Small, Fl. S.E.U.S. 1339. 1903.
Solidago rigida L. Sp. Pl. 2:880. 1753.

Perennial from a stout, thickened caudex and short-creeping rhizomes; stems 1 to several, stout, erect to ascending, to 1.5 m tall, densely hispid, becoming glabrate near the base, or rarely completely glabrous, longitudinally striate; basal leaves usually present at flowering time, thick or thin, firm or rarely membranous, elliptic to broadly lanceolate to broadly ovate, acute to obtuse at the apex, tapering to a long petiole, entire or serrate, gray-pubescent on both surfaces, pinnately nerved, up to 25 cm long, up to 8 cm wide; middle and upper cauline leaves progressively smaller, lanceolate to elliptic, usually entire, sessile or nearly so; inflorescence a round- or flat-topped corymb with 30–200 showy heads, the branches open or contracted; involucre 5–9 mm high; phyllaries in 3–5 series, unequal, firm, oblong, obtuse or acute, glabrous or strigillose, striate; receptacle epaleate; ray flowers 6–12, 3.5–7.5 mm long, pistillate, yellow; disc flowers up to 35, 4.5–6.0 mm long, bisexual, yellow; cypselae obovoid, glabrous or pubescent only at the tip, 10- to 15-nerved, 1–2 mm long; pappus of numerous white capillary bristles slightly thickened at the tip, 3.0–5.5 mm long.

Two varieties occur in Illinois:

a. Outer phyllaries strigillose; leaves and stems hispid or strigose; branches of the inflorescence open . . . . 4a. O. rigidum var. rigidum

a. Outer phyllaries glabrous; leaves and stems glabrous or sparsely hirsute; branches of the inflorescence contracted. . . . . 4b. *O. rigidum* var. *glabratum*

4a. **Oligoneuron rigidum** (L.) Small var. **rigidum**. Fig. 45.

Outer phyllaries strigillose; inner phyllaries glabrous to sparsely strigillose, oblong, obtuse; branches of the inflorescence open; leaves and stems coarsely hispid.

Common Name: Stiff goldenrod.
Habitat: Dry prairies.
Range: Massachusetts to Ontario and South Dakota, south to Texas, Missouri, and South Carolina.
Illinois Distribution: Common in Illinois, except for the southernmost counties.

This is the common variety of *O. rigidum* in Illinois. It flowers from July to October.

4b. **Oligoneuron rigidum** (L.) Small var. **glabratum** (E. L. Braun) G. L. Nesom, Phytologia 75:27. 1993.
Solidago rigida L. var. *glabrata* E. L. Braun, Rhodora 44:3–4. 1942.
Solidago rigida L. ssp. *glabrata* (E. L. Braun) S. B. Heard & Semple, Can. Journ. Bot. 66:1807. 1988.

Outer phyllaries glabrous; leaves and stems glabrous or sparsely hirsute; branches of the inflorescence contracted.

Common Name: Smooth stiff goldenrod.
Habitat: Dry prairies.
Range: Ohio to Missouri, south to Texas and Georgia.
Illinois Distribution: Very rare; Jackson Co.: DeSoto Railroad Prairie, apparently extirpated.

This variety flowers from August to October.

45. *Oligoneuron rigidum* (Stiff goldenrod).
a. Upper part of plant.
b. Lower leaf.
c. Flowering head.
d. Phyllary.
e. Ray flower.
f, g. Disc flowers.
h. Cypsela.

5. **Oligoneuron X lutescens** (Lindl. ex DC.) G. L. Nesom, Phytologia 75:29. 1992. Fig. 46.

Diplopappus lutescens Lindl. ex DC. Prodr. 5:278. 1834.
Aster lutescens (Lindl. ex DC.) Hook. ex Torr. & Gray, Fl. N. Am. 2:160. 1884.
Aster ptarmicoides Torr. & Gray var. *lutescens* (Lindl. ex DC.) Gray, Syn. Fl. N. Am. 1:199. 1884.

Perennial from a branched, thickened caudex; stems 1 to a few, erect to ascending, to 50 cm tall, glabrous or less commonly hirtellous; previous year's leaves often persistent at base of plant; lower leaves of the current season linear to broadly lanceolate, firm, acute to obtuse at the apex, tapering to the sessile base, entire or nearly so, glabrous but often scabrous, pinnately nerved, up to 20 cm long, up to 2 cm wide; upper leaves smaller, linear; inflorescence a flat-topped corymb with 2–50 heads on strongly ascending branches; involucre 5–8 mm high, nearly as wide, turbinate; phyllaries in 3 or 4 series, unequal, oblong, firm, acute to obtuse, greenish, glabrous; receptacle epaleate; ray flowers 8–20, 5–8 mm long, pistillate, cream or yellowish; disc flowers 15–20, 4–5 mm long, bisexual, white or yellowish; cypselae obovoid, flat, glabrous, several-nerved, 1.2–1.6 mm long; pappus of numerous capillary bristles enlarged at the tip, 3.5–5.5 mm long.

Common Name: Yellow stiff aster.
Habitat: Dry prairies.
Range: Ontario to Manitoba, south to South Dakota, Illinois, and Indiana.
Illinois Distribution: Rare; known only from Cook County.

This is considered to be a hybrid between *O. album* and *O. riddellii*.

This hybrid flowers during August and September.

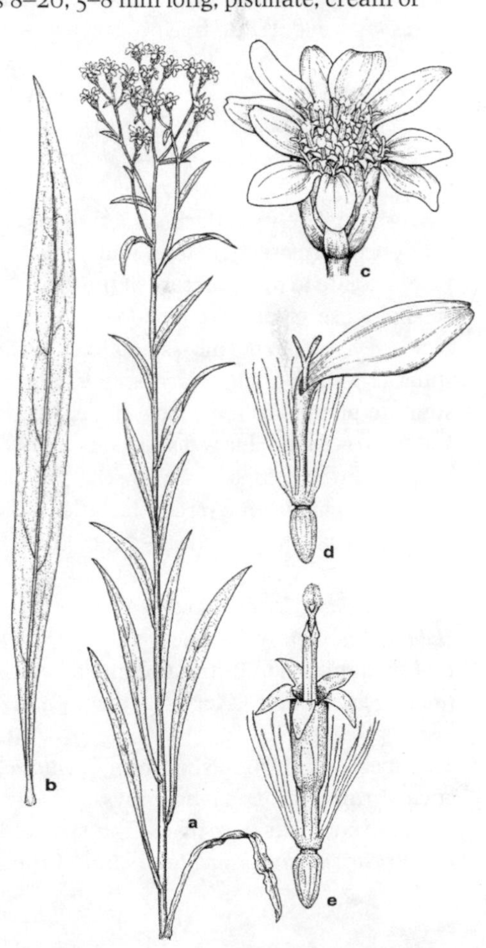

46. *Oligoneuron X lutescens* (Stiff yellow aster).
a. Upper part of plant.
b. Leaf.
c. Flowering head.
d. Ray flower.
e. Disc flower.

10. **Chrysopsis** Ell.—Golden Aster

Perennial herbs, sometimes with rhizomes; stems branched, sometimes pubescent, sometimes resinous; cauline leaves alternate, sessile, entire or serrate; heads many, radiate; involucre hemispheric; receptacle flat, epaleate; phyllaries in 4 or 5 series, unequal; ray flowers yellow, pistillate; disc flowers numerous, yellow, tubular, perfect; cypselae flat, pubescent; pappus of ray flowers and disc flowers double, the inner of capillary bristles, the outer of scales or short, stiff bristles.

Chrysopsis, in my opinion, contains several species in the United States and Mexico. Some botanists restrict *Chrysopsis* to the southeastern United States, placing the other species in *Heterotheca*. The cypselae of both the ray flowers and the disc flowers are flat. In *Heterotheca*, the cypselae of the disc flowers are thickened, while those of the ray flowers are flat.

1. Leaves with few to several teeth; plants with rhizomes .... 1. *C. camporum*
1. Leaves entire or nearly so; plants without rhizomes...................... 2. *C. villosa*

1. **Chrysopsis camporum** Greene, Pittonia 3:88. 1897. Fig. 47.

Chrysopsis villosa (Pursh) Nutt. var. *camporum* (Greene) Cronq. Bull. Torrey Club 74:150. 1947.
Heterotheca camporum (Greene) Shinners, Field & Lab. 19:71. 1951.
Heterotheca villosa (Pursh) Shinners var. *camporum* (Greene) Wunderlin, Ann. Mo. Bot. Gard. 59:471. 1973.

Perennial herbs with rhizomes; stems ascending to erect, to 75 cm tall, branched, strigose to hirsute, sometimes stipitate-glandular; upper cauline leaves oblanceolate to ovate, obtuse at the apex, cuneate at the sessile base, sparsely serrate, strigose, ciliate, to 7 cm long, to 1.5 cm wide; lower cauline leaves petiolate; heads several, in corymbs, sometimes borne single, to 2 cm across; involucre hemispheric, 7–10 mm high; phyllaries in 4 or 5 series, unequal, strigose, sometimes stipitate-glandular; receptacle flat, epaleate; ray flowers up to 30, pistillate, golden yellow, 10–20 mm long; disc flowers up to 60, bisexual, tubular, yellow, 5–6 mm high, glabrous or nearly so; cypselae flat, obovate, 1.5–4.0 mm long, strigose, with 7–10 ribs; pappus in 2 series, the outer of linear scales up to 1 mm long, the inner of whitish capillary bristles 5–7 mm long.

Common Name: Prairie golden aster.
Habitat: Sandy prairies.
Range: Manitoba to British Columbia, south to New Mexico and Alabama.
Illinois Distribution: Occasional throughout Illinois.

This species is distinguished from *C. villosa* by its sparsely serrate leaves, the presence of rhizomes, and longer rays.

Some botanists have placed this species in *Heterotheca*.

Chrysopsis camporum flowers from June to September.

47. *Chrysopsis camporum*
(Prairie golden aster).
a. Upper part of plant.
b. Middle stem and leaves.
c. Flowering head.
d. Phyllary.
e. Ray flower.
f. Disc flower.
g. Cypsela.
h. Upper part of cypsela.

2. **Chrysopsis villosa** (Pursh) Nutt. ex DC. Prodr. 5:327. 1836. Fig. 48.
Chrysopsis villosa (Pursh) Nutt. var. *minor* Hook. London Journ. Bot. 6:244. 1847.
Heterotheca villosa (Pursh) Shinners, Field & Lab. 19:71. 1951.
Heterotheca villosa (Pursh) Shinners var. *minor* (Hook.) Semple, Novon 4:54. 1994.

48. *Chrysopsis villosa* (Hairy prairie golden aster).
a. Upper part of plant.
b. Flowering head.
c. Phyllary.
d. Ray flower.
e. Disc flower.
f. Cypsela.

Perennial herbs without rhizomes; stems ascending to erect, to 30 cm tall, strigose or sparsely hirsute; lower cauline leaves broadly oblanceolate to oblong, obtuse to acute at the apex, tapering to the sometimes rounded, sessile base, entire or nearly so, strigose, glandular, to 8 cm long, to 2 cm wide; heads several, in corymbs, sometimes solitary; involucre cylindric, 6.5–8.5 mm high; phyllaries in 4 or 5 series, unequal, narrowly lanceolate, long-strigose; receptacle flat, epaleate; ray flowers up to 25, pistillate, golden yellow, 6–12 mm long; disc flowers up to 50, bisexual, tubular, yellow, 5–6 mm high, glabrous or nearly so; cypselae flat, obovate. 1.5–4.0 mm long, strigose, with 7–10 ribs; pappus in 2 series, the outer of linear scales up to 1 mm long, the inner of whitish capillary bristles 5–7 mm long.

Common Name: Hairy prairie golden aster.
Habitat: Sandy soil.
Range: Wisconsin to Washington, south to California, Texas, and Illinois.
Illinois Distribution: Scattered in Illinois.

This species is sometimes placed in the genus *Heterotheca*.

Our plants are sometimes considered to be var. *minor*, a taxon that differs by its broadly oblanceolate to oblong lower cauline leaves, its more sparse pubescence, and it long-strigose phyllaries.

Chrysopsis villosa flowers from June to September.

11. **Heterotheca** Cass.—Camphorweed

Aromatic annual or perennial herbs; stems ascending to erect, pubescent; basal leaves tapering to the base; cauline leaves alternate, auriculate at the base; heads several, radiate, in corymbs or borne singly; involucre hemispheric; phyllaries in 4–6 series, linear to narrowly lanceolate, unequal; receptacle flat, epaleate; ray flowers up to 30, golden yellow, pistillate; disc flowers up to 10, tubular, bisexual, golden yellow, perfect or staminate; cypselae of ray flowers thickened, of disc flowers flat, pubescent; pappus of ray flowers of a few bristles or absent, of disc flowers of two types, the inner a row of capillary bristles, the outer a row of shorter stouter bristles or scales.

Six species, as considered here, comprise the genus.

1. **Heterotheca subaxillaris** (Lam.) Britt. & Rusby, Trans. N.Y. Acad. Sci. 7:10. 1887. Fig. 49.

Inula subaxillaris Lam. Encycl. 3:259. 1799.
Heterotheca latifolia Buckl. Proc. Acad. Nat. Sci. Phila. 13:459. 1863.
Heterotheca subaxillaris (Lam.) Britt. & Rusby ssp. *latifolia* (Buckl.) Semple, Sida 21:759. 2004.

Annual or biennial herbs; stems erect, to 1 m tall, hirsute to hispid-strigose, often stipitate-glandular; basal leaves oblong to ovate, acute at the apex, tapering to a long-petiolate base, dentate to serrate, to 8 cm long, to 5 cm wide; cauline leaves alternate, lanceolate to elliptic to ovate, acute at the apex, tapering to the sessile or clasping base, to 5 cm long, to 3 cm wide, hispid-strigose, often scabrous, sometimes stipitate-glandular; heads several to numerous, in corymbs or panicles, radiate, to 1.5 cm across; involucre hemispheric to campanulate, 4–8 mm high; phyllaries in 4–6 series, linear to lanceolate, unequal, hispid to strigose, stipitate-glandular; receptacle flat, epaleate; ray flowers 15–35, pistillate, golden yellow, to 9 mm long; disc flowers up to 60, bisexual or staminate, tubular, golden yellow, to 9 mm high; cypselae of ray flowers thickened, of disc flowers flat, 1.5–2.5 mm long, glabrous or occasionally strigose; pappus of ray flowers absent or of a few caducous capillary bristles, of disc flowers an inner series of capillary bristles and an outer row of short, stout bristles or scales.

49. *Heterotheca subaxillaris* (Camphorweed).

a. Upper part of plant.
b. Leaf.
c. Flowering head.
d. Phyllary.
e. Ray flower.
f. Cypsela of ray flower.
g. Disc flower.
h. Cypsela of disc flower.

Common Name: Camphorweed; golden aster.
Habitat: Sandy soil, disturbed soil.
Range: Throughout the United States.
Illinois Distribution: Occasional in southern Illinois where it is native; in disturbed soil in northern Illinois where it is introduced.

Heterotheca subaxillaris is similar to our two species of *Chrysopsis* but differs by having two kinds of cypselae and by characteristics of the pappus.

Some botanists recognize two subspecies of *H. subaxillaris*, calling ours ssp. *latifolia*. I am not convinced that these two are that distinct.

This species flowers from July to October.

12. **Erigeron** L.—Fleabane

Annual, biennial, or perennial herbs with fibrous roots or rhizomes; stems erect to ascending, branched or unbranched, glabrous or pubescent; leaves basal or cauline and alternate, or both, sessile or petiolate, sometimes clasping, usually 1-nerved, entire to toothed (in Illinois) to pinnatifid, glabrous or pubescent; heads single or in corymbs or panicles, radiate; involucre hemispheric or turbinate; phyllaries in 3–5 series, equal or unequal; receptacle flat or conic, pitted, epaleate; ray flowers usually 50 to numerous, pistillate, white or pink; disc flowers numerous, bisexual, yellow, tubular, 5-lobed; cypselae oblong to oblong-obovoid, flat and 2- or 4-nerved (in Illinois), glabrous or pubescent; pappus of outer scales or setae to 0.4 mm long and inner stramineous barbellate bristles.

Nearly 400 species comprise the genus, most of them in temperate regions of the world.

Key to the Species of *Erigeron* in Illinois

1. Leaves broadly rounded at the sessile or nearly sessile or clasping base.
    2. Heads few, 2.5–3.5 cm across; rays 50–100, about 1 mm wide....... 1. *E. pulchellus*
    2. Heads several, 1.5–2.0 cm across; rays 150–200, about 0.5 mm wide... 2. *E. philadelphicus*
1. Some or all the leaves petiolate, not clasping, sometimes reduced upwards.
    3. Plants 0.3–1.5 m tall or taller; blades of lower leaves 2.5–15.0 cm long; pappus of ray flowers absent or simple.
        4. Cauline leaves many; basal leaves ovate, coarsely dentate; at least the middle of the stem with long, spreading hairs .... 3. *E. annuus*
        4. Cauline leaves few; basal leaves spatulate, entire or nearly so; middle part of stem with short, appressed hairs ....4. *E. strigosus*
    3. Plants up to 0.4 m tall; blades of lower leaves 1–3 cm long; pappus of all flowers double .... 5. *E. tenuis*

1. **Erigeron pulchellus** Michx. Fl. Bor. Am. 2:124. 1803. Fig. 50.
Erigeron bellidifolius Muhl. ex Willd. Sp. Pl. 3:1958. 1804.

Perennial herbs from spreading rhizomes; stems usually unbranched, erect, soft, hollow, to 60 (–75) cm tall, villous to hirsute; leaves of two types: basal leaves in a rosette, obovate to spatulate, obtuse at the apex, tapering to the very short-petiolate base, to 10 cm long, to 4 cm wide, crenate to dentate, villous to hirsute; cauline leaves remote, alternate, lanceolate to elliptic to narrowly ovate, acute at the apex, tapering to the sessile or sometimes clasping base, entire or serrulate, villous to hirsute or less commonly nearly glabrous, the lowest the largest; flowering heads 1–7, 2.5–3.5 cm across, on villous peduncles; involucre hemispheric, 5–7 mm high; phyllaries in 2 or 3 series, more or less equal, linear, acuminate at the apex, villous to hirsute; receptacle flat, pitted, epaleate; ray flowers 50–100, about 1 mm wide, pistillate, usually pink or lavender, 6–10 mm long; disc flowers numerous, tubular, bisexual, yellow, 3.5–6.0 mm high; cypselae flattened, usually glabrous or sparsely strigose, 1.3–1.8 mm long, 2- or 4-nerved; pappus of numerous capillary bristles in one series.

50. *Erigeron pulchellus* (Robin's plantain).

a. Habit.
b. Rosette of basal leaves.
c. Flowering head, face view.
d. Involucre.
e. Phyllary.
f. Ray flower.
g. Disc flower.
h. Cypsela.

Common Name: Robin's plantain.
Habitat: Open woods, dry shaded banks.
Range: Quebec to Minnesota, south to Texas and Florida.
Illinois Distribution: Occasional throughout Illinois.

This handsome species has the largest flowering heads in the genus in Illinois. Some or all of the cauline leaves may be subclasping but not as prominent as in *E. philadelphicus*. Most Illinois botanists until the twenty-first century called this species *E. bellidifolius*.

Erigeron pulchellus flowers from April to mid-June.

2. **Erigeron philadelphicus** L. Sp. Pl. 2:863. 1753. Fig. 51.

Annual, biennial, or perennial herbs from short rhizomes; stems erect, slender, branched above, to 1 m tall, hirsute to villous to nearly glabrous, at least above, minutely glandular; leaves of 2 types: basal leaves several in a rosette, obovate to spatulate, obtuse to subacute at the apex, tapering to the short-petiolate base, dentate, to 10 cm long, to 3 cm wide, hirsute or villous or commonly nearly glabrous; cauline leaves alternate, remote, the lowest ones spatulate to obovate, obtuse to acute at the apex, dentate, the uppermost elliptic to lanceolate, acute at the apex, tapering to the usually clasping base, to 3 cm long, to 1.5 cm wide, dentate to serrate to entire, usually short-pubescent; flowering heads up to 35, 1.5–2.0 cm across, on short-hairy peduncles; involucre more or less hemispheric, 4–6 mm high; phyllaries in 2 or 3 series, usually about equal, linear, green with scarious margin, spreading-hirsute to glabrous; receptacle more or less flat, pitted, epaleate; ray flowers 150–200, about 0.5 mm wide, pistillate, white or sometimes pink, 5–10 mm long; disc flowers numerous, tubular, bisexual, yellow, 2.0–3.5 mm high; cypselae flattened, sparsely strigose, 2-nerved, 0.5–1.0 mm long; pappus of white capillary bristles in one series.

Common Name: Philadelphia fleabane.
Habitat: Mesic woods, wet meadows, dry prairies, fields, along rivers and streams, disturbed areas.
Range: Labrador to British Columbia, south to California, Texas, and Florida.
Illinois Distribution: Common throughout the state.

This species is readily distinguished from the other species of *Erigeron* in Illinois by its distinctly clasping cauline leaves. The flowering heads are smaller and more numerous than those of *E. pulchellus*.

The rays of most specimens are white, but pink-rayed plants are relatively common.

Erigeron philadelphicus is common in wetlands and in rich mesophytic woods throughout the state.

The flowers may bloom in early March and continue nearly to July.

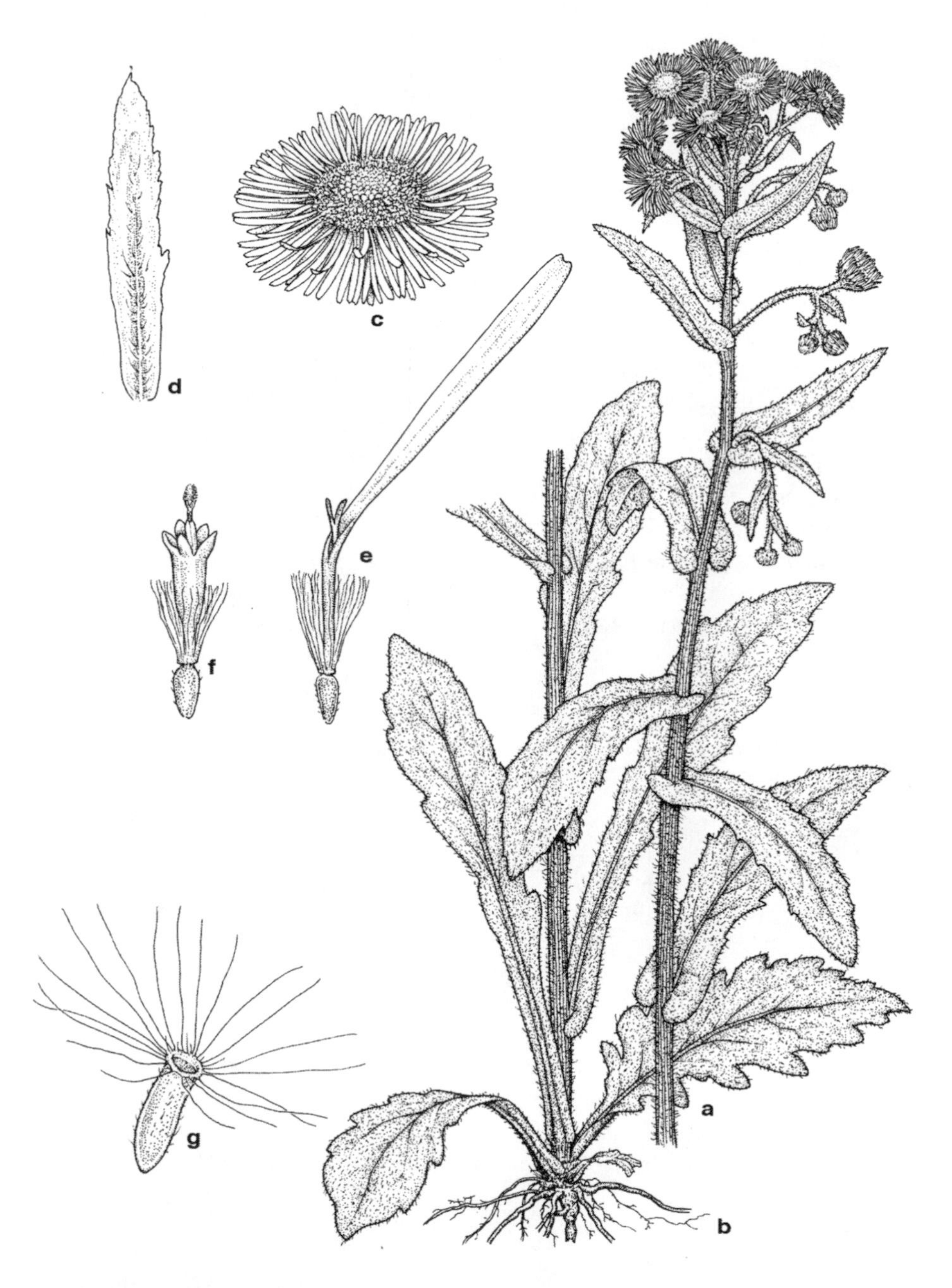

51. *Erigeron philadelphicus* (Philadelphia fleabane).
a. Upper part of plant.
b. Lower part of plant.
c. Flowering head.
d. Phyllary.
e. Ray flower.
f. Disc flower.
g. Cypsela.

3. **Erigeron annuus** (L.) Pers. Syn. Pl. 2:431. 1807. Fig. 52.
Aster annuus L. Sp. Pl. 2:875. 1753.

Annual herbs from fibrous roots; stems erect, usually branched, to 1.5 m tall, spreading-hispid, sometimes strigose near the top, eglandular; leaves of two kinds: basal leaves withered at anthesis, ovate to ovate-lanceolate, obtuse at the apex, tapering to the marginate petiolate base, to 10 cm long, to 5 cm wide, dentate, usually hispid; cauline leaves alternate, numerous, lanceolate to oblong, acute to subacute at the apex, tapering to the sessile base, to 3 cm long, to 1 cm wide, serrate or the uppermost sometimes entire, hispid; heads several to numerous in a branched panicle, radiate, to 1.2 cm across, on slender, hispid peduncles; involucre hemispheric, up to 5 mm

52. *Erigeron annuus* (Eastern daisy fleabane).

a. Habit.
b. Upper part of plant.
c. Lower leaf.
d. Flowering head.
e. Phyllary.
f. Ray flower.
g. Disc flower.
h. Cypsela.

high; phyllaries in 2–4 series, more or less equal, linear, villous to hirsute, minutely glandular; receptacle flat, pitted, epaleate; ray flowers up to 125, pistillate, white, less commonly pink, 4–10 mm long; disc flowers numerous, bisexual, tubular, yellow, 2–3 mm long; cypselae flat, 0.8–1.0 mm long, 2-nerved, strigose; pappus in 1 or 2 series, the inner series of white capillary bristles, the outer, when present, scalelike.

Common Names: Eastern daisy fleabane; annual fleabane; white-top.
Habitat: Old fields, roadsides, pastures, disturbed areas, including woods.
Range: Newfoundland to British Columbia, south to California, Texas, and Florida.
Illinois Distribution: Common throughout the state; in every county.

This common plant of disturbed areas and old fields differs from the similar-appearing *E. strigosus* by its sharply dentate leaves and its spreading-hispid stems.

The rays are sometimes pink. The pappus may be in 1 or 2 series, and may be bristlelike or scalelike.

This species usually flowers from May to September but has been collected in flower in Illinois in early November.

4. **Erigeron strigosus** Muhl. ex Willd. Sp. Pl. 3:1956. 1803. Fig. 53.
Stenactis beyrichii Fisch. & Mey. Index Sem. Hort. Petrop. 5. 1838.
Erigeron strigosus Muhl. ex Willd. var. *beyrichii* (Fisch. & Mey.) Torr. & Gray, Syn. Fl. N. Am. 1:219. 1884.
Erigeron ramosus (Walt.) BSP. Prel. Cat. N.Y. 1888.
Erigeron ramosus (Walt.) BSP. var. *beyrichii* (Fisch. & Mey.) Trel. Rep. Ark. Geol. Survey 1888:192. 1891.
Erigeron annuus (L.) ssp. *strigosus* (Muhl. ex Willd.) Wageritz, Ill. Fl. Mittel-Eur., ed. 2, 6/3: 96. 1965.

Usually an annual herb with fibrous roots; stems erect, often branched, strigose or rarely hirsute, to 1 m tall, eglandular; leaves of 2 kinds: basal leaves often persistent at anthesis, spatulate to oblanceolate to oblong, obtuse to acute at the apex, tapering to a slender emarginate petiole, to 15 cm long, to 2 cm wide, strigillose; cauline leaves alternate, few, linear-lanceolate to narrowly oblong, acute at the apex, tapering to the sessile base, entire or sparingly serrate, strigillose to glabrous, to 3 cm long, to 1 cm wide; heads usually numerous, up to 1.2 cm across, radiate, in panicles; involucre (2–) 3–4 mm high; phyllaries in 2–4 series, more or less equal, linear, glabrous to strigose; receptacle flat, pitted, epaleate; ray flowers 50–100, pistillate, usually white but occasionally pinkish, 4–6 mm long; disc flowers numerous, yellow, bisexual, tubular, 1.5–2.5 mm high; cypselae flat, 0.7–1.2 mm long, 2-nerve, strigose; pappus in 2 series, the inner scalelike, the outer bristlelike.

Common Name: Common eastern fleabane.
Habitat: Old fields, roadsides, dry woods, dry prairies, disturbed areas.
Range: Nova Scotia to Washington, south to Texas and Florida.
Illinois Distribution: Common throughout the state; in every county.

Erigeron strigosus differs from *E. annuus* by its strigose stems and its usually entire cauline leaves. Plants with the involucre 2–3 mm high occur in Illinois and may be called var. *beyrichii*. Several botanists in Illinois until 1940 called this species *Erigeron ramosus*, but this is not the *Erigeron ramosus* of Rafinesque (1817).

This species flowers from late May through September.

53. *Erigeron strigosus* (Common eastern fleabane).
a. Upper part of plant.
b. Middle stem and leaves.
c. Flowering head.
d. Phyllaries.
e. Ray flower.
f. Disc flowers.
g. Cypsela.

5. **Erigeron tenuis** Torr. & Gray, Fl. N. Am. 2:175. 184. Fig. 54.

Annual or biennial herb from fibrous roots; stems erect to ascending, branched from the base, usually strigose, up to 40 cm tall; leaves of 2 kinds: basal leaves usually persistent at anthesis, oblanceolate to obovate, obtuse to acute at the apex, tapering to the petiolate base, sparsely hirsute, to 3 cm long (in Illinois), to 1 cm wide, serrate (in Illinois); cauline leaves alternate, linear, acute at the apex, tapering to the sessile base, sparsely hirsute to glabrous, to 2 cm long, to 0.8 cm wide, entire; heads 1–20, to 1.8 cm across, radiate; involucre 2–6 mm high; phyllaries in 2–4 series, mostly equal, linear, sparsely strigose to glabrous; receptacle flat, pitted, epaleate; ray flowers 60–100 (–120), pistillate, white (in Illinois) or pale blue, 3–5 mm long; disc flowers numerous, bisexual, yellow, tubular, 2.0–2.5 mm high; cypselae

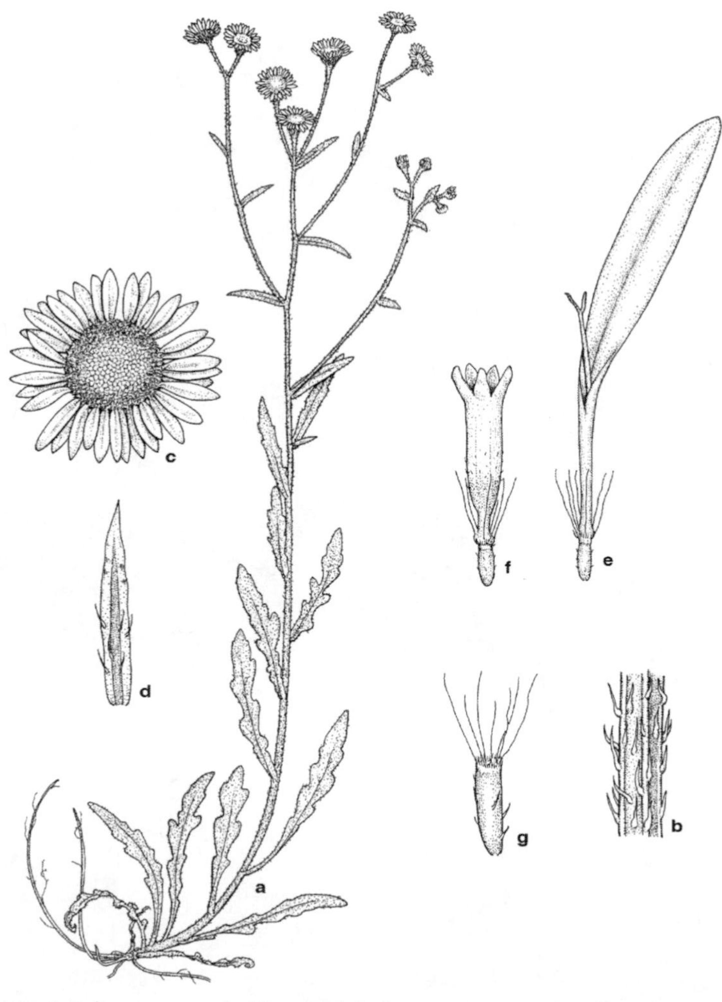

54. *Erigeron tenuis* (Slender-leaf fleabane). a. Habit. b. Stem with hairs. c. Flowering head. d. Phyllary. e. Ray flower. f. Disc flower. g. Cypsela.

flat, 1.0–1.2 mm long, usually 2-nerved, sparsely strigose; pappus in 2 distinct series, the outer of small setae, the inner of capillary bristles.

Common Name: Slender-leaf fleabane.
Habitat: Roadside (in Illinois).
Range: Missouri to Kansas, south to Texas and Mississippi; adventive in Illinois.
Illinois Distribution: Along a road in Union County.

This adventive species in Illinois is much shorter than *E. annuus* and *E. strigosus* but has two distinct rows of pappus; the outer row is composed of setae, the inner of capillary bristles.

This species flowers in May and June.

13. **Conyza** Less., *nom. conserv.*—Horseweed

Annual or biennial herbs; leaves alternate, simple; inflorescence a raceme or thyrse or panicle of small heads, radiate; involucre campanulate; phyllaries in 2 or 3 series, usually unequal; receptacle epaleate; ray flowers short, white to purplish, pistillate; disc flowers tubular, bisexual, the corollas usually 4-lobed; cypselae flat; pappus of capillary bristles in 1 series.

Many years ago this genus was considered to be a part of *Erigeron*, although in 1818, Rafinesque had separated it into his genus *Leptilon*. Although Rafinesque's *Leptilon* was named several years before Lessing named *Conyza*, *Conyza* has been conserved. *Conyza* differs from *Erigeron* by its short, campanulate involucre and its smaller ray and disc flowers. Recent botanists have placed the species of *Conyza* back into *Erigeron*.

There are approximately 20 species native to North America and Asia.

1. Stems unbranched or nearly so up to the inflorescence; rays white; plants to 3.5 m tall. . . . . 1. *C. canadensis*
1. Stems diffusely branched from near the base; rays purplish; plants rarely more than 25 cm tall. . . . . 2. *C. ramosissima*

1. **Conyza canadensis** (L.) Cronq. Bull. Torrey Club 70:632. 1943. Fig. 55.
Erigeron canadensis L. Sp. Pl. 2:863. 1753.
Leptilon canadense (L.) Britt. in Britt. & Brown, Ill. Fl. N.E.U.S. 3:391. 1898.

Annual or biennial herb with fibrous roots or a somewhat thickened caudex; stems erect, unbranched below, usually several-branched near the inflorescence, to 3.5 m tall, hispid to hirsute, rarely nearly glabrous; leaves alternate, crowded, the lower spatulate to oblanceolate, the upper linear, acute at the apex, tapering to the usually sessile base, the lower to 7 (–10) cm long, to 1 cm wide, the upper up to 1.5 cm long, to 0.8 cm wide, all of them hirsute to hispid and usually ciliolate; heads numerous in racemes or cymes or panicles, up to 4 mm across, radiate; involucre campanulate, 2.5–4.5 mm high; phyllaries in 2 or 3 series, the outer linear to narrowly lanceolate, glabrous to strigose, the inner linear and shorter than the outer phyllaries; receptacle epaleate; ray flowers up to 50, white, 0.5–1.0 mm long, pistillate; disc flowers up to 30, tubular, bisexual, yellowish, 0.5–0.8 mm high; cypselae flat, tan or gray, 1.0–1.5 mm long, sparsely strigose; pappus of up to 25 white capillary bristles 2–3 mm long.

Common Name: Horseweed; mare's-tail.
Habitat: Old fields, pastures, disturbed areas, often weedy.
Range: Nova Scotia to British Columbia, south to California, Texas, and Illinois; Mexico; Central America.
Illinois Distribution: Common throughout the state; in every county.

Although this species is common in disturbed areas, it is considered to be native. Plants may flower when only a few centimeters tall to nearly 3.5 m tall.

The stem is usually unbranched below, but often has numerous ascending branches just below the inflorescence.

Conyza canadensis flowers from May to November.

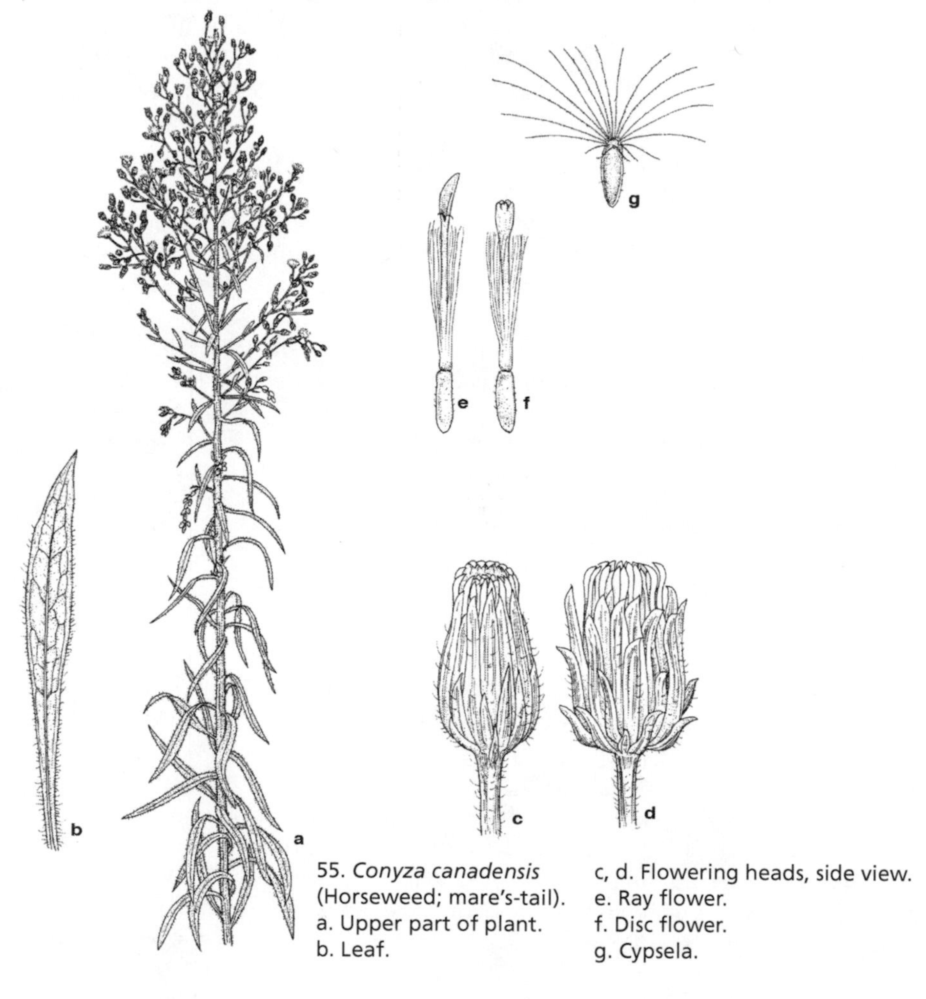

55. *Conyza canadensis* (Horseweed; mare's-tail). a. Upper part of plant. b. Leaf. c, d. Flowering heads, side view. e. Ray flower. f. Disc flower. g. Cypsela.

2. **Conyza ramosissima** Cronq. Bull. Torrey Club 70:632. 1943. Fig. 56.
Erigeron divaricatus Michx. Fl. Bor. Am. 2:123. 1803, *non Conyza divaricata* Sprengel (1826).
Leptilon divaricatum (Michx.) Raf. Am. Monthly Mag. 268. 1818.

Low-growing annual herb with fibrous roots; stems branched from the base, to 30 cm tall, strigose; leaves alternate, linear or even subulate, entire, to 15 mm long, to 2.0 (–2.5) mm wide, strigose to hirsute; heads numerous in corymbs or borne singly, 2–3 mm across, radiate; involucre campanulate, 2–4 mm high; phyllaries in 2 series, the outer lanceolate, green to purplish, strigillose, the inner linear, stramineous to purplish, usually glabrous; receptacle epaleate; ray flowers up to 30, pistillate, purplish, up to 0.8 mm long; disc flowers up to 8, tubular, bisexual, yellowish, 0.3–0.6 mm high; cypselae flat, tan, 1.0–1.5 mm long; pappus of capillary bristles, tawny to pinkish, 2.0–2.5 mm long.

Common Name: Dwarf fleabane.
Habitat: Disturbed soil, particularly sandy areas.
Range: Wisconsin to North Dakota, south to Texas and Mississippi.
Illinois Distribution: Occasional throughout the state.

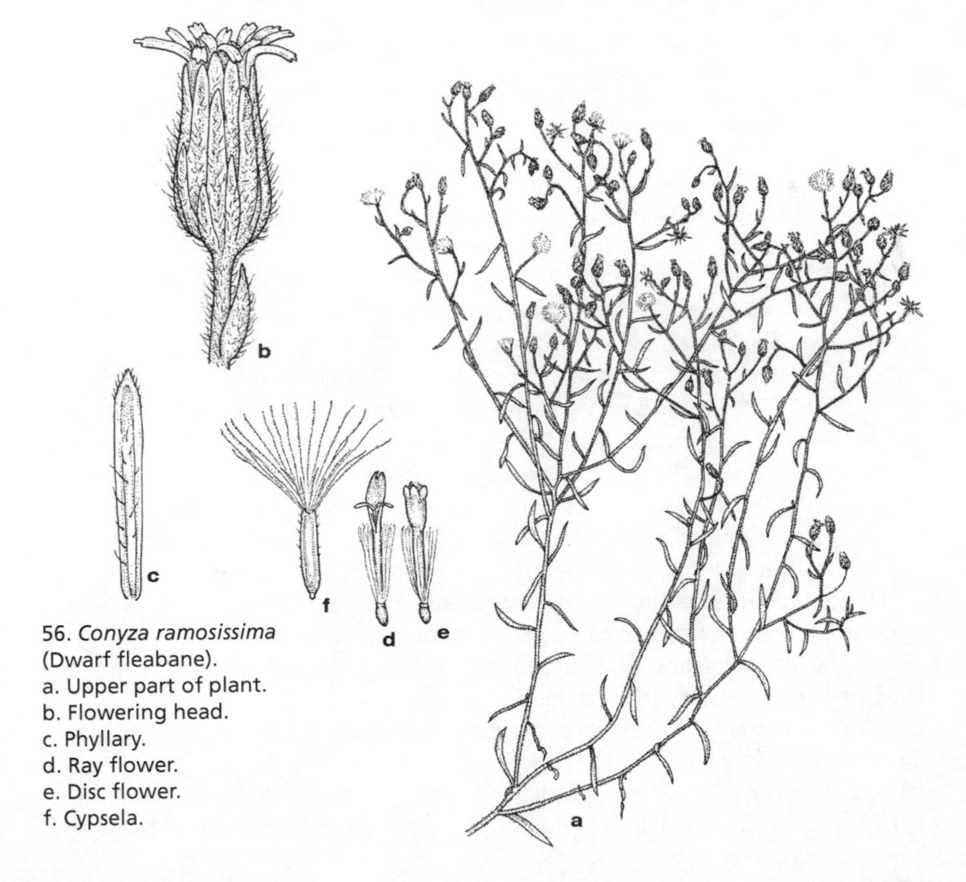

56. *Conyza ramosissima* (Dwarf fleabane).
a. Upper part of plant.
b. Flowering head.
c. Phyllary.
d. Ray flower.
e. Disc flower.
f. Cypsela.

This diminutive species is distinguished by its much-branched stems, linear leaves, and tiny flowering heads with purplish rays.

Although Michaux first named this plant *Erigeron divaricatus*, when transferred to *Conyza* it cannot become *Conyza divaricata* because Sprengel had used that binomial in 1826 for a different species.

Conyza ramosissima flowers from June to September.

14. **Boltonia** L'Her.—Doll's-daisy; False Aster

Perennial herbs, usually with stolons; stems erect to ascending, branched, striate; leaves of 2 types: basal leaves withered at flowering time, the cauline alternate, all leaves entire and glabrous; heads several to many in corymbs or panicles; involucre hemispheric; phyllaries in 3–6 series, equal or unequal; receptacle flat, epaleate; ray flowers pistillate, usually white; disc flowers bisexual, yellow, tubular, 5-lobed; cypselae from ray flowers trigonous and narrowly winged, from disc flowers compressed with or without wings; pappus in 1 series, of 2 or 3 awns and 2–4 short, soft bristles.

Most species of *Boltonia* resemble *Symphyotrichum*, differing by the lack of capillary bristles on the cypselae.

There are five species in the genus, all in North America.

Key to the Species of *Boltonia* in Illinois

1. Leaves linear to oblanceolate, 5–20 mm wide; disc 6–10 mm wide.
   2. Leaves not decurrent . . . . 1. *B. asteroides*
   2. Leaves decurrent . . . . 2. *B. decurrens*
1. Leaves linear, 1–5 mm wide; disc 3–6 mm wide . . . . 3. *B. diffusa*

1. **Boltonia asteroides** (L.) L'Her. Sert. Angl. 27. 1789.
Matricaria asteroides L. Mant. Pl. 116. 1767.

Perennial herbs with stolons; stems erect, corrugated, glabrous, up to 1.5 m tall; cauline leaves alternate, linear to lanceolate, acute at the apex, tapering to the sessile base, 1-nerved, entire, glabrous, usually with a bluish tint, to 20 cm long, to 2 cm wide; heads in corymbs or panicles on spreading to ascending branches, on leafy peduncles up to 20 cm long, radiate; involucre hemispheric, 3–5 mm high; phyllaries in 3–5 series, subulate to linear to spatulate to oblanceolate, the lowest merging on to the peduncle, the outer larger than the inner; receptacle flat, epaleate; ray flowers up to 60, usually white but sometimes lilac, pistillate, 5–13 mm long; disc flowers numerous, bisexual, yellow, tubular, 1.5–2.5 mm high, the disc 6–10 mm wide; cypselae obovoid, 1–3 mm long, very narrowly winged; pappus of stiff awns up to 2 mm long.

There is variation within this species in characteristics of the phyllaries. Some botanists in the past have recognized three separate species, while others consider them to be three varieties. I have mixed feelings about this, but I have elected, at least for now, to recognize three varieties.

Variety *asteroides* is not known from Illinois, although I suspect that it will be found eventually in the state. The typical variety has subulate to linear phyllaries with acute apices. The other two varieties, both known from Illinois, have narrowly oblanceolate to oblanceolate to spatulate to obovate phyllaries with cuspidate apices.

The following key separates the two varieties of *Boltonia asteroides* in Illinois:

a. Phyllaries obovate to spatulate, with membranous margins 2.5–6.0 mm wide; rays 8–10 mm long. . . . . .1a. *B. asteroides* var. *latisquama*

a. Phyllaries narrowly oblanceolate to oblanceolate, with membranous margins 1.0–2.5 mm wide; rays 10–15 mm long . . . . 1b. *B. asteroides* var. *recognita*

1a. **Boltonia asteroides** (L.) L'Her. var. **latisquama** (Gray) Cronq. Bull. Torrey Club 74:149. 1947. Fig. 57.

Boltonia latisquama Gray, Am. Journ. Sci. & Arts, ser. 2, 33:238. 1862.

Phyllaries obovate to spatulate, more or less equal, with membranous margins 2.5–6.0 mm wide, cuspidate at the apex; rays 8–10 mm long; pappus awns 0.7–2.0 mm long.

Common Name:
White doll's-daisy; false aster.

Habitat: Moist ground.

Range: Maine to North Dakota, south to Oklahoma, Mississippi, and Virginia.

Illinois Distribution:
Known only from Cook County.

This variety flowers from May to October.

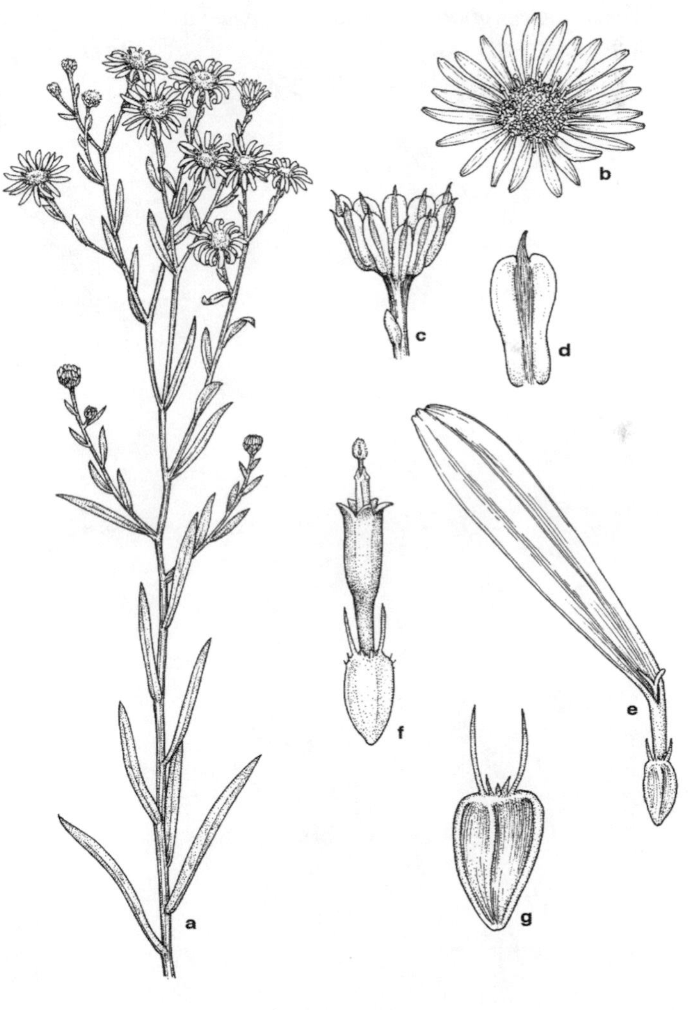

57. *Boltonia asteroides* var. *latisquama* (White doll's-daisy).
a. Upper part of plant.
b. Flowering head.
c. Involucre.
d. Phyllary.
e. Ray flower.
f. Disc flower.
g. Cypsela.

1b. **Boltonia asteroides** (L.) L'Her. var. **recognita** (Fern. & Grisc.) Cronq. Bull. Torrey Club 74:149. 1947. Fig. 57.

Boltonia latisquama Gray var. *recognita* Fern. & Grisc. Rhodora 42:491. 1940.
Boltonia latisquama Gray var. *microcephala* Fern. & Grisc. Rhodora 42:492. 1940.
Boltonia recognita (Fern. & Grisc.) G. N. Jones, Vasc. Plants of Ill. 428. 1955.

Phyllaries narrowly oblanceolate to oblanceolate, with membranous margins 1.0–2.5 mm wide; rays 10–15 mm long.

Common Name: White doll's-daisy; false aster.
Habitat: Moist ground, marshes, prairies, sometimes in standing water.
Range: Maine to Oregon, south to Oklahoma, Kentucky, Tennessee, and New Jersey.
Illinois Distribution: Common throughout the state.

This is the common variety of *B. asteroides* in Illinois. It occurs in almost any wetland habitat. Its leaves often have a bluish cast to them.

This variety flowers from July to October.

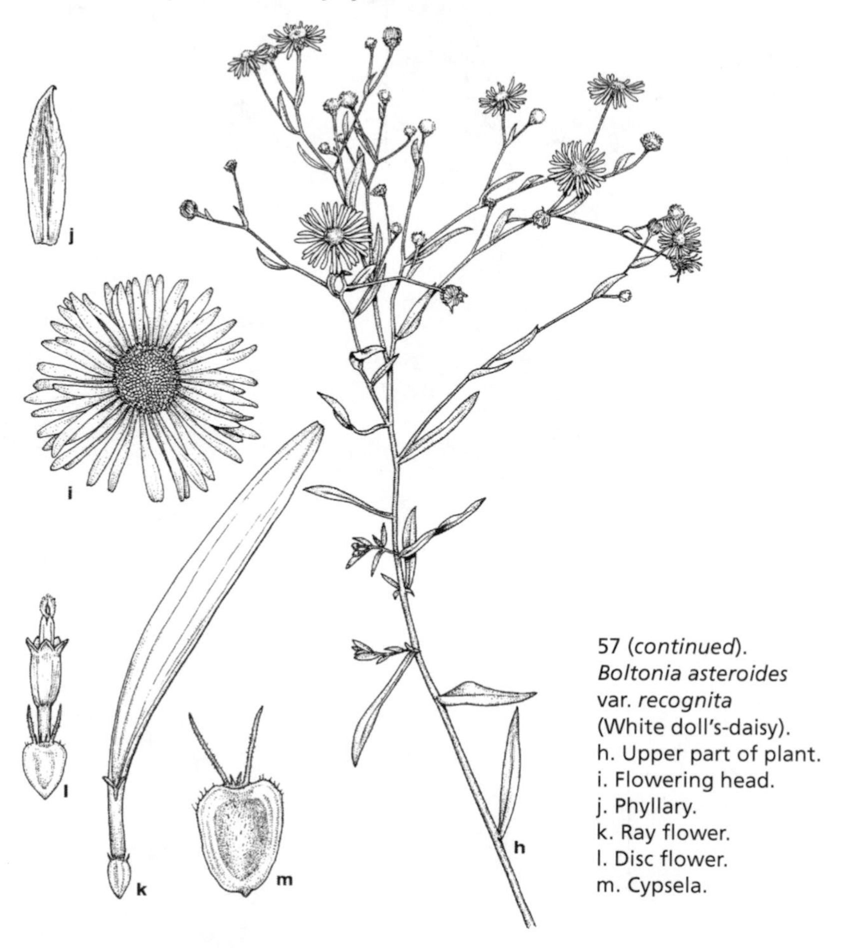

57 (*continued*). *Boltonia asteroides* var. *recognita* (White doll's-daisy). h. Upper part of plant. i. Flowering head. j. Phyllary. k. Ray flower. l. Disc flower. m. Cypsela.

2. **Boltonia decurrens** (Torr. & Gray) A. Wood, Am. Bot. Fl. 166. 1870. Fig. 58.
Boltonia glastifolia Michx. var. *decurrens* Torr. & Gray, Fl. N. Am. 2:188. 1842.
Boltonia asteroides (L.) L'Her. var. *decurrens* (Torr. & Gray) Engelmann ex Gray, Syn. Fl. N. Am. 1:166. 1884.
Boltonia latisquama Gray var. *decurrens* (Torr. & Gray) Fern. & Grisc. Rhodora 42:492. 1940.

Perennial herb with fibrous roots; stems erect, glabrous, up to 2 m tall; cauline leaves alternate, linear-lanceolate to lanceolate, acute at the apex, tapering to the decurrent and sagittate base, 1-nerved, entire, glabrous, to 15 cm long, to 2 cm wide; heads few to several in corymbs or panicles, on peduncles up to 5 cm long, radiate; involucre hemispheric, 3.0–4.5 mm high; phyllaries in 3–5 series, oblong to spatulate, more or less equal; receptacle flat, epaleate; ray flowers up to 60, pistillate, white or sometimes lilac, 10–15 mm long; disc flowers numerous, bisexual, yellow, tubular, 1.8–2.5 mm high, the disc 6–10 mm wide; cypselae obovoid, 1.5–2.5 mm long, very narrowly winged; pappus of stiff awns up to 1.8 mm long.

Common Name: Clasping-leaf doll's-daisy; decurrent false aster.
Habitat: Moist ground, mostly along or near the Illinois River.
Range: Western Illinois and eastern Missouri.
Illinois Distribution: Confined to counties along the Illinois River; adventive in Lake Co.

Except for the decurrent leaves, this species differs very little from *B. asteroides* var. *recognita*. Some botanists have considered it to be a variety of *B. asteroides*.

Boltonia decurrens is on the list of Federally Threatened Species.

This species flowers from July to October.

58. *Boltonia decurrens* (Clasping-leaf doll's-daisy).
a, b. Upper part of plant.
c. Stem with decurrent leaves.
d. Flowering head.
e. Ray flower.
f. Disc flower.
g. Cypsela.

3. **Boltonia diffusa** Ell. Sketch Bot. S. Carol. 2:400. 1823.

Perennial herbs, with or without stolons; stems erect, slender, much branched, glabrous, to 1.5 (–2.0) m tall; cauline leaves alternate, linear to linear-lanceolate, acute at the apex, tapering to the sessile base, 1-nerved, entire, glabrous, to 10 cm long, to 5 mm wide; heads numerous in panicles with spreading branches, on thickened or filiform peduncles 3–15 cm long, radiate; involucre hemispheric, 2.5–3.0 mm high; phyllaries in 4–6 series, subulate to linear-oblong, 0.2–0.5 mm wide, more or less equal; receptacle flat, epaleate; ray flowers up to 40, pistillate, white or lilac, 3–7 mm long; disc flowers numerous, bisexual, tubular, yellow, 1.2–2.5 mm high, the disc 3–6 mm wide; cypselae obovoid, 1.5–2.5 mm long, very narrowly winged; pappus of stiff awns less than 1 mm long.

This species differs from both *B. asteroides* and *B. decurrens* by its smaller flowering heads and its narrower leaves.

There is variation in the shape of the phyllaries, the thickness of the peduncles, and the presence or absence of stolons. Two varieties may be recognized, distinguished by the following key:

a. Phyllaries subulate; peduncles filiform; plants with stolons . . . . 1. *B. diffusa* var. *diffusa*
a. Phyllaries linear-oblong; peduncles thickened; plants without stolons. . . . . 2. *B. diffusa* var. *interior*

3a. **Boltonia diffusa** Ell. var. **diffusa**. Fig. 59.

Plants with slender stolons; phyllaries subulate; peduncles filiform.

Common Name: Small-head doll's-daisy; narrow-leaved false aster.
Habitat: Moist ground.
Range: North Carolina to Missouri, south to Texas and Florida.
Illinois Distribution: Rare; known only from Union County.

This variety flowers from July to September.

3b. **Boltonia diffusa** Ell. var. **interior** Fern. & Grisc. Rhodora 42:290. 1940. Fig. 59.

Boltonia interior (Fern. & Grisc.) G. N. Jones, Vasc. Plants of Ill. 478. 1955.

Plants not stoloniferous; phyllaries linear-oblong; peduncles thickened, not filiform.

Common Name: Narrow-leaved false aster.
Habitat: Moist or dry open ground.
Range: Arkansas, Illinois, Kentucky, Mississippi, Missouri, Oklahoma, Tennessee.
Illinois Distribution: Occasional in the southern half of the state, north to Marion County.

This is the more common variety of *B. diffusa* in Illinois. The characteristics that distinguish var. *interior* from the typical variety may be sufficient enough to merit species status. It flowers from July to September.

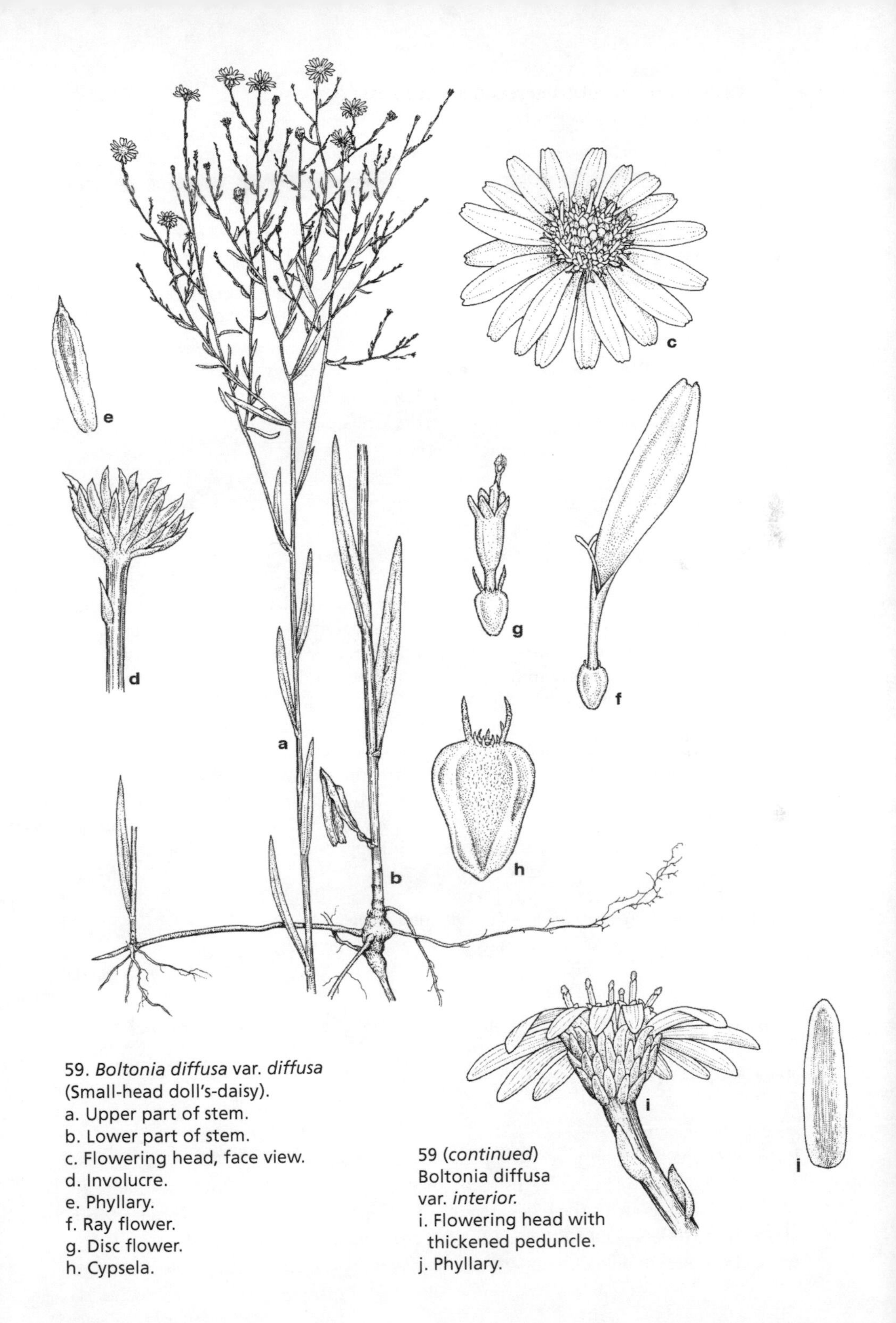

59. *Boltonia diffusa* var. *diffusa* (Small-head doll's-daisy).
a. Upper part of stem.
b. Lower part of stem.
c. Flowering head, face view.
d. Involucre.
e. Phyllary.
f. Ray flower.
g. Disc flower.
h. Cypsela.

59 (*continued*)
Boltonia diffusa
var. *interior.*
i. Flowering head with thickened peduncle.
j. Phyllary.

15. **Eurybia** S. F. Gray—Big-leaf Aster

Perennial herbs with rhizomes; stems 1 to several, erect to ascending, glabrous or pubescent; basal leaves present at flowering time, cordate (in Illinois), long-petiolate; cauline leaves alternate, progressively smaller toward the top of the stem; heads radiate, in a corymb; involucre campanulate; phyllaries in 3–7 series, unequal, linear to ovate, 1-nerved, glabrous or pubescent; receptacle more or less flat, pitted, epaleate; ray flowers white to purple, pistillate; disc flowers yellow, tubular, bisexual; cypselae obconic to fusiform, more or less flat, several-nerved, glabrous or pubescent; pappus of reddish to yellowish barbellate capillary bristles somewhat thickened at the tip in 2–4 series.

This genus, often considered a part of the traditional *Aster* in the past, has persistent, cordate basal leaves (in Illinois) on long petioles and pappus with reddish to yellowish capillary bristles that are somewhat thickened at the tip.

There are 23 species in this genus, all in North America.

Key to the Species of *Eurybia* in Illinois

1. Peduncles and involucre not stipitate-glandular.
    2. Leaves scabrous on the upper surface, with prominent venation; basal rosettes absent in spring.
        3. Leaves more or less glabrous beneath, rather thin, with fragile petioles; branches of inflorescence compact; cordate base of leaves strongly overlapping; heads about 2.5 cm across; involucre 8–10 mm high, narrowly campanulate; ray flowers about 8, 10–12 mm long; disc flowers about 25, red-brown; pappus bristles red-brown, 4–5 mm long . . . . 1. *E. chasei*
        3. Leaves spreading-hirsute beneath, thick, with rigid petioles; branches of inflorescence widely forking; cordate base of leaves not overlapping; heads up to 2 cm across; involucre 6–8 mm high, campanulate; ray flowers 9–20, 10–18 mm long; disc flowers up to 50, yellow; pappus bristles yellowish, 6–7 mm long . . . . 2. *E. furcata*
    2. Leaves not at all or sparingly scabrous on the upper surface, without prominent venation; basal rosettes present in spring . . . . 3. *E. schreberi*
1. Peduncles and involucre stipitate-glandular.
    4. Rays white; lower surface of leaves villous and eglandular. . . . . 3. *E. schreberi*
    4. Rays purple; lower surface of leaves strigose and stipitate-glandular . . . . 4. *E. macrophylla*

1. **Eurybia chasei** (G. N. Jones) Mohlenbr. Guide Vasc. Fl. Ill., ed. 4, 126. 2013. Fig. 60.

Aster chasei G. N. Jones, Vasc. Fl. Ill. 466. 1955.

Perennial from long, creeping rhizomes, sometimes forming colonies; stems erect, to 60 cm tall, glabrous or puberulent; basal leaves absent at flowering time, not forming a rosette in the spring, thin, ovate, acute to subacuminate at the apex, cordate at the base with the lobes strongly overlapping, serrate, somewhat scabrous above, more or less glabrous beneath, to 15 cm long, to 15 cm wide, on fragile, winged petioles; middle and upper cauline leaves progressively smaller, lance-ovate to lance-elliptic, serrate, on short-winged petioles, or the uppermost sessile; heads about 2.5 cm across, 20–40 in a flat- or round-topped corymb, the branches ascending and compact, not widely forking, not stipitate-glandular; involucre 8–10 mm high, longer than wide, narrowly campanulate, not stipitate-glandular; phyllaries in several series, unequal, linear to narrowly lanceolate, obtuse, with short, green tips, coriaceous, scarious, ciliolate, sparsely pubescent; receptacle flat, pitted, epaleate; ray flowers about 8, 10–12 mm long, white, or lavender when old, pistillate; disc flowers about 25, red-brown, tubular, bisexual, 6–8 mm high; cypselae not flat, linear to narrowly oblongoid, green-brown, 2–5 mm long, striate, glabrous; pappus of firm reddish bristles 4–5 mm long.

Common Name: Chase's aster.
Habitat: Moist or wet, shaded ravines.
Range: West-central Illinois.
Illinois Distribution: Only known from Marshall, Peoria, and Tazewell counties.

This Illinois endemic is apparently confined to moist or wet wooded slopes in three counties in west-central Illinois. It was first found in September 1902 by F. E. McDonald in Peoria County. The plant is named for Virginius Chase, a noted amateur botanist, who also was familiar with the plant and also collected it in Peoria County.

Eurybia chasei differs from *E. furcata* in several ways, as indicated in the key above and in the discussion under *E. furcata* below.

This species flowers in August and September.

60. *Eurybia chasei* (Chase's aster).
a. Upper part of plant.
b. Middle of stem with leaves.
c. Flowering head.
d. Phyllaries.
e. Ray flower.
f. Disc flower.
g. Cypsela.

2. **Eurybia furcata** (E. S. Burgess) G. L. Nesom, Phytologia 77:259. 1995. Fig. 61.
Aster furcatus E. S. Burgess in Britt. & Brown, Ill. Fl. N. U. S. 3:358. 1898.
Aster furcatus E. S. Burgess in Britt. var. *elaciniatus* Benke, Am. Midl. Nat. 13:326. 1932.

Perennial from long, creeping rhizomes, often forming colonies; stems erect, often slightly zigzag, to 1.2 m tall, puberulent above, usually glabrous below; basal leaves absent at flowering time, not forming a rosette in the spring, thick, firm, ovate to ovate-lanceolate, acuminate at the apex, cordate at the base, serrate, hispidulous or hirsute and scabrous on both surfaces, to 15 cm long, to 8 cm wide, on firm, winged, laciniate petioles, the basal lobes not overlapping; middle and upper cauline leaves progressively smaller, lance-ovate to lance-elliptic, serrate, on short winged petioles, or the uppermost sessile; heads up to 2 cm across, 10–25 in a flat- or round-topped corymb, the branches ascending, widely forking, not stipitate-glandular; involucre 6–8 mm high, about as wide, campanulate, not stipitate-glandular; phyllaries in several series, unequal, linear to narrowly lanceolate, obtuse, with short green tips, usually puberulent, sometimes purplish; receptacle flat, pitted, epaleate; ray flowers 9–20, 10–18 mm long, white, or lavender when old, pistillate; disc flowers up to 50, tubular, yellow, bisexual, 6–8 mm high; cypselae not flat, linear to narrowly oblongoid, greenish, 2.5–4.0 mm long, 8- to 12-nerved, sparsely pubescent; pappus of firm yellowish bristles 6–7 mm long.

Common Name: Forked aster.
Habitat: Wooded bluffs, wooded slopes, often calcareous areas.
Range: Michigan to Wisconsin, south to Iowa, Arkansas, Illinois, and Indiana.
Illinois Distribution: Uncommon in the northeastern corner of Illinois; also Tazewell County.

A plant without laciniate petioles was found at Crystal Lake in McHenry County on September 2, 1928, by H. C. Benke and was named var. *elaciniatus* by Benke.

Eurybia furcata and *E. chasei* differ from the other two species of *Eurybia* in Illinois by the absence of glandular-pubescence on the branches of the inflorescence. *Eurybia furcata* also differs from *E. chasei* by its more pubescent leaves, its widely forking inflorescence, its smaller flowering heads, its shorter and broader involucre, its more numerous ray and disc flowers, and its yellowish pappus bristles 6–7 mm long. The basal lobes of the leaves are strongly overlapping in *E. chasei*.

Eurybia furcata flowers in August and September.

3. **Eurybia schreberi** (Nees) Nees, Gen. Sp. Aster. 137. 1832. Fig. 62.
Aster schreberi Nees, Syn. Aster. Herb. 16. 1818.

Perennial herbs from creeping rhizomes and often a thickened caudex, colonial; stem solitary, erect, to 1.2 m tall, glabrous to spreading-hairy or puberulent, but not glandular; tufts of large leaves present at base of plant on short sterile shoots; basal leaves thin, ovate, acuminate at the apex, cordate at the base, serrate-dentate,

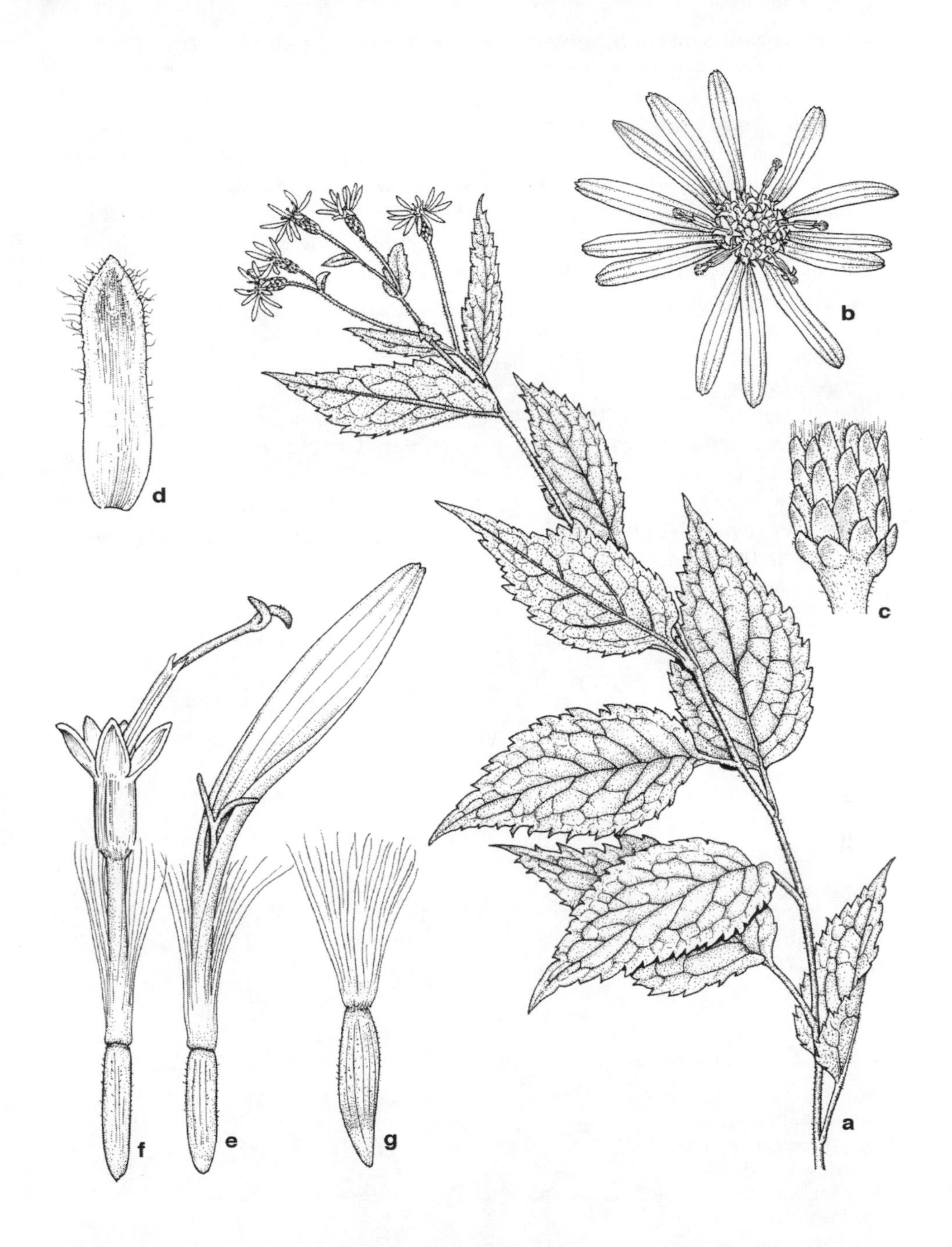

61. *Eurybia furcata* (Forked aster).
a. Upper part of plant.
b. Flowering head, face view.
c. Involucre.
d. Phyllary.
e. Ray flower.
f. Disc flower.
g. Cypsela.

glabrous to strigillose to hirsute on both surfaces, up to 20 cm long, up to 15 cm wide, on wingless petioles up to 10 cm long; cauline leaves alternate, narrowly ovate to broadly lanceolate, acuminate at the apex, cordate or rounded at the base, sometimes sessile, to 7 cm long, to 3 cm wide; heads 15–100 in flat- or round-topped corymbs; involucre 5–8 mm high, 3–8 mm wide, sometimes stipitate-glandular; phyllaries in 4 or 5 series, unequal, oblong to lanceolate, glabrous, ciliate, eglandular; receptacle flat, pitted, epaleate; ray flowers 6–14, 10–12 mm long, pistillate, white; disc flowers up to 30, bisexual, 5–7 mm long, yellow; cypselae fusiform, flat, linear, 3.2–3.7 mm long, with 6–12 nerves, glabrous or strigillose; pappus of orange bristles 5–7 mm long.

Common Name: Schreber's aster.
Habitat: Woods.
Range: Maine to Ontario to Wisconsin, south to Iowa, Illinois, Indiana, and Virginia.
Illinois Distribution: Rare; scattered in the northern third of the state.

This is the only *Eurybia* in Illinois with white rays that do not turn lavender with age. It also differs from *E. macrophylla* by its eglandular phyllaries and from *E. furcata* by its winged petioles.

Eurybia schreberi flowers from July to September.

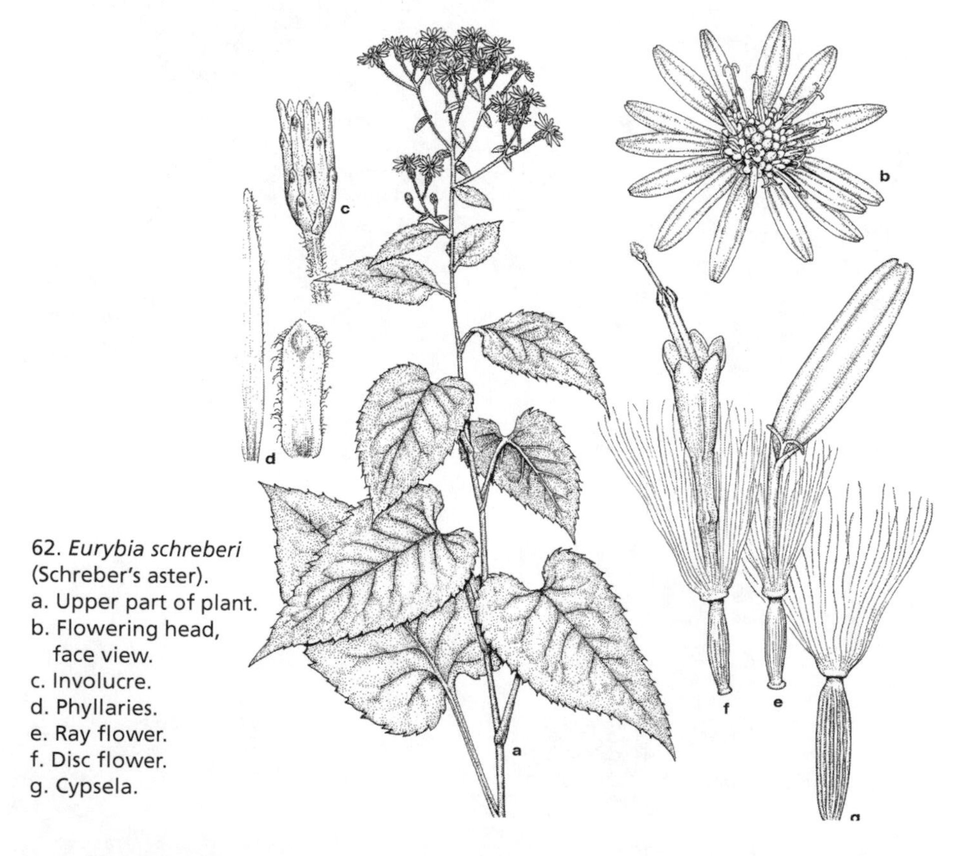

62. *Eurybia schreberi* (Schreber's aster).
a. Upper part of plant.
b. Flowering head, face view.
c. Involucre.
d. Phyllaries.
e. Ray flower.
f. Disc flower.
g. Cypsela.

4. **Eurybia macrophylla** (L.) Cass. in F. Cuvier, Dict. Sci. Nat., ed. 2, 37:487. 1825. Fig. 63.

Aster macrophyllus L. Sp. Pl., ed. 2, 2:1232. 1763.

Aster ianthinus E. S. Burgess in Britt. Ill. Fl. N.U. S. 3:360. 1828.

Aster macrophyllus L. var. *ianthinus* (E. S. Burgess) Fern. Fl. Southington 29. 1902.

63. *Eurybia macrophylla* (Big-leaf aster).

a. Habit.
b. Lower leaf.
c. Flowering head.
d. Involucre.
e. Phyllary.
f. Ray flower.
g. Disc flowers.
h. Cypsela.

Perennial herbs from long, slender, fleshy creeping rhizomes and often a thickened caudex, forming colonies; stems erect, often slightly zigzag, to 1.5 m tall, spreading-hairy, the hairs stipitate-glandular, at least in the upper part; tufts of large leaves present at base of plant on short sterile shoots; lowest leaves thick or rarely thin, firm, ovate, acuminate at the apex, cordate at the base, crenate to sharply serrate, glabrous to hirsute on both surfaces, the hairs often gland-tipped, to 25 cm long, to 15 cm wide, on winged petioles up to 12 cm long; cauline leaves alternate, progressively smaller, on winged petioles, narrowly ovate to lance-elliptic, acute at the apex, tapering to the base, glabrous or scabrous above, strigose or stipitate-glandular below, to 10 cm long, to 5 cm wide; heads 15–25 in a flat- or round-topped corymb with gland-tipped hairs on the branches; involucre 7–11 mm high, about as wide, stipitate-glandular; phyllaries in 5–7 series, unequal, linear to linear-lanceolate, 4.5–6.0 mm long, 1–2 mm wide, the outer ovate, obtuse, stipitate-glandular and often short-hairy; receptacle flat, pitted, epaleate; ray flowers 10–20, 7–15 mm long, pistillate, lavender or purple, rarely white with a purplish base; disc flowers up to 40 (–50), 6–8 mm long, bisexual, tubular, yellow; cypselae narrowly oblongoid, more or less flat, 7- to 12-nerved, glabrous or sparsely pubescent; pappus of yellowish or reddish, firm bristles 7–9 mm long.

Common Name: Big-leaf aster.
Habitat: Dry open woods, slopes of wooded dunes, swampy forests.
Range: New Brunswick to Manitoba, south to Missouri, Tennessee, and Georgia.
Illinois Distribution: Confined to the northernmost counties of Boone, Cook, DuPage, and Lake.

This species differs by its purple rays and stipitate glands on the lower surface of the leaves.

Eurybia macrophylla flowers from July to October.

16. **Machaeranthera** Nees—Tadoka Daisy

Annual or biennial herbs from a taproot; leaves basal and cauline, the basal withered at flowering time, the cauline alternate, pinnatifid; heads several, radiate; involucre hemispheric; phyllaries in 3–7 series, stipitate-glandular; receptacle flat to convex, epaleate; ray flowers up to 50, blue or purple, pistillate, fertile; disc flowers up to 150, tubular, yellow, 5-lobed, bisexual, fertile; cypselae obovoid, several-ribbed, pubescent; pappus of numerous white or tawny bristles in 1–3 series.

Two species comprise this genus. Both at one time were placed in the genus *Aster.* Only the following species has been found in Illinois.

1. **Machaeranthera tanacetifolia** (Kunth) Nees, Gen. Sp. Aster. 225. 1832. Fig. 64.
Aster tanacetifolia Kunth in HBK. Nov. Gen. Sp. 4:95. 1820.

Annual or biennial herb from a taproot; stems erect, much branched, glandular-pubescent, viscid, to 85 cm tall; leaves 2- or 3-pinnatifid, to 10 cm long, to 1.2 cm wide, the divisions linear to oblong, pubescent, the lowest petiolate, the upper sessile;

heads numerous, up to 3.5 cm across, in corymbs or panicles; involucre hemispheric, 8–10 mm high; phyllaries in 5–7 series, linear, acute at the apex, appressed to spreading to reflexed, glandular; receptacle more or less flat, epaleate; ray flowers up to 25, pistillate, purple, up to 20 mm long; disc flowers up to 150, tubular, yellow, bisexual, 5-lobed, 4–7 mm high; cypselae obovoid, several-ribbed, pubescent, 2–4 mm long; pappus of numerous tawny capillary bristles to 8 mm long.

64. *Machaeranthera tanacetifolia* (Tadoka daisy).
a. Upper part of plant.
b. Leaf.
c. Clusters of fruiting heads.
d, e. Flowering heads.
f. Phyllary.
g. Ray flower.
h. Disc flower.
i. Cypsela.

Common Name: Tadoka daisy; tansy aster.
Habitat: Disturbed soil.
Range: Native to the plains of the United States; introduced in Illinois.
Illinois Distribution: Escaped from cultivation in Jackson County.

This plant has blue rays similar to those found in some species of *Symphyotrichum*, but the pinnatifid leaves are totally different from any species of *Symphyotrichum*.

Machaeranthera tanacetifolia flowers in August and September.

17. **Prionopsis** Nutt.

Annual or biennial herb; leaves alternate, simple, spinulose-dentate; heads several, in corymbs, radiate; involucre hemispheric; phyllaries in 3–5 series, somewhat resinous; receptacle more or less flat, epaleate; ray flowers yellow, pistillate; disc flowers yellow, tubular, bisexual; cypselae flat, glabrous; pappus of numerous persistent capillary bristles.

Only the following species comprises the genus.

1. **Prionopsis ciliata** (Nutt.) Nutt. Trans. Am. Phil. Soc. 7:329. 1841. Fig. 65.
Donia ciliata Nutt. Journ. Acad. Nat. Sci. Phila. 2:118. 1821.
Grindelia ciliata (Nutt.) Spreng. Syst. Veg. 3:575. 1826.
Haplopappus ciliata (Nutt.) DC. Prodr. 5:346. 1836.

Annual or perennial herb; stems erect, branched, glabrous, to 1.5 m tall; cauline leaves alternate, oblong to obovate, obtuse at the apex, more or less rounded at the sessile and usually clasping base, to 6 cm long, to 3 cm wide, sharply spinulose-dentate; heads several in corymbs or borne singly, radiate, to 3 cm across; involucre hemispheric, 10–15 mm high; phyllaries in 3–5 series, linear, spreading to appressed, somewhat resinous; receptacle more or less flat, epaleate; ray flowers up to 45, 10–15 mm long, golden yellow, pistillate; disc flowers numerous, tubular, golden yellow, bisexual; cypselae gray, flat, glabrous, 2–4 mm long; pappus of 15 or more persistent capillary bristles 7–10 mm long.

Common Name: Golden aster; prionopsis.
Habitat: Sandy soil along roadsides (in Illinois).
Range: Michigan to California, east to Louisiana.
Illinois Distribution: Known only from a roadside in Alexander County.

Our species of *Prionopsis*, *Chrysopsis*, *Heterotheca*, and *Grindelia* are strikingly similar in appearance with their yellow or golden yellow rays. *Prionopsis* is clearly distinct by the pappus, which consists only of 15 or more persistent capillary bristles.

Some botanists place this species in *Grindelia*, a concept not followed here because of the pappus characteristics.

Prionopsis ciliata flowers in August and September.

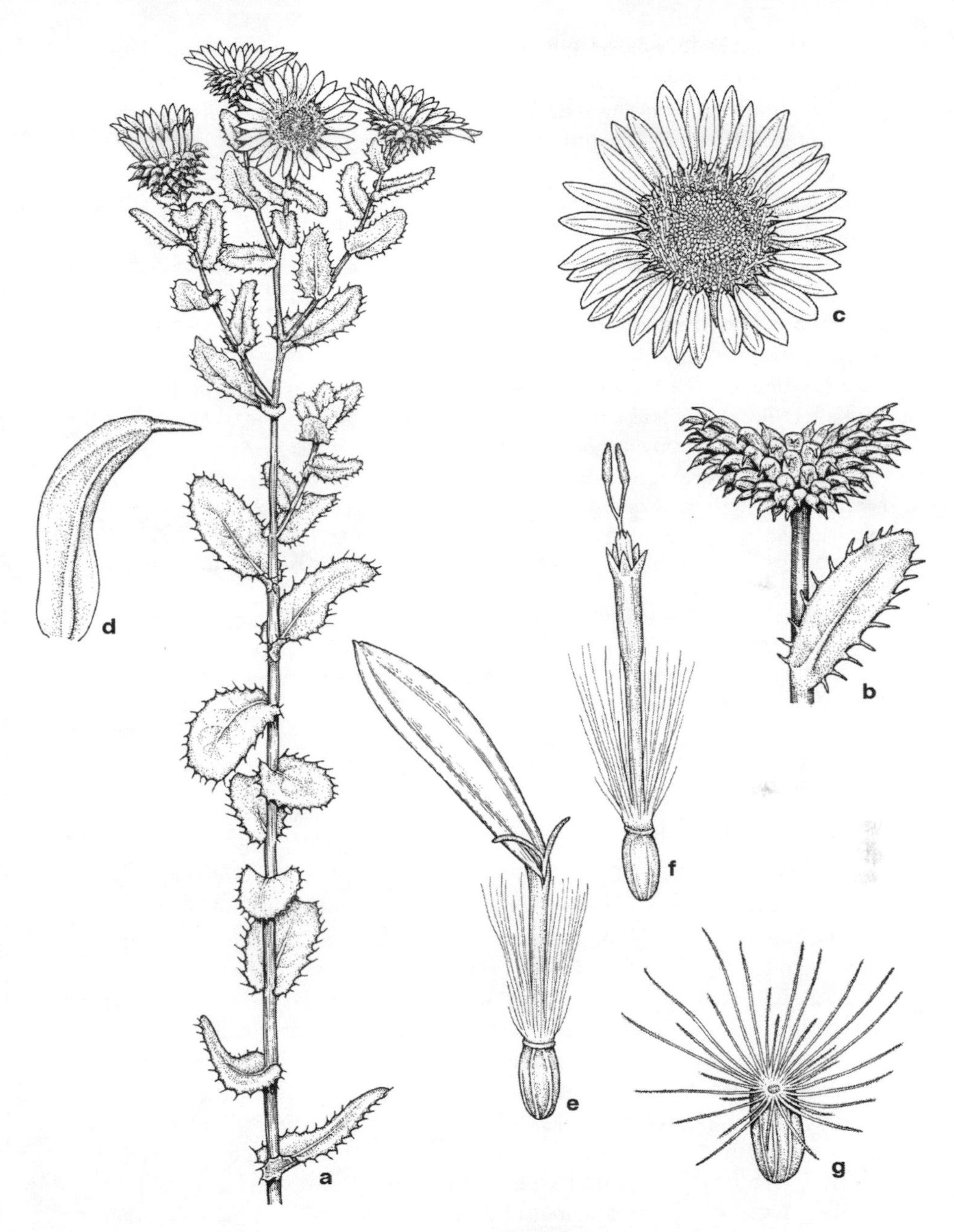

65. *Prionopsis ciliata*
(Golden aster).
a. Upper part of plant.
b. Lower part of involucre and bract.
c. Flowering head.
d. Phyllary.
e. Ray flower.
f. Disc flower.
g. Cypsela.

18. **Grindelia** Willd.—Gum-plant

Annual or perennial herbs (in Illinois) or subshrubs; stems erect or less commonly prostrate, branched or unbranched, often resinous; basal leaves petiolate; cauline leaves alternate, 1-nerved, serrate to crenate, sessile; heads radiate or discoid, in corymbs or panicles; involucre hemispheric to campanulate to globose; phyllaries in 3–9 series, linear to lanceolate, more or less unequal, glabrous, more or less resinous; receptacle pitted, epaleate; ray flowers up to 60, rarely absent, yellow, pistillate, fertile; disc flowers numerous, tubular, yellow, bisexual, fertile, less commonly only staminate; cypselae flat or thickened, often 3- or 5-angled, glabrous; pappus of 2–8 deciduous awns, scales, or bristles in 1 series.

This genus of about 30 species in the western United States, Mexico, and South America has a complex taxonomy. Some botanists include *G. ciliata* in this genus, but I am maintaining it as a member of the genus *Prionopsis*, partly because of the pappus of numerous capillary bristles that tend to be persistent.

I recognize three species in Illinois, separated by the following key:

1. Annuals; leaves serrulate to crenulate; phyllaries squarrose; cypselae 2–3 mm long . . . . 1. *G. squarrosa*
1. Perennials; leaves with bristle-tipped teeth; phyllaries squarrose or not squarrose; cypselae 4–6 (–7) mm long.
    2. Pappus of two setiform awns 4–8 mm long, equaling or longer than the disc corolla; phyllaries not squarrose . . . . 2. *G. lanceolata*
    2. Pappus of 2–6 setiform awns or scales 2–5 mm long, usually shorter than the disc corolla; phyllaries squarrose . . . . 3. *G. perennis*

1. **Grindelia squarrosa** (Pursh) Dunal, Mem. Mus. Hist. Nat. 5:50. 1819.
Donia squarrosa Pursh, Fl. Am. Sept. 559. 1814.

Biennial or perennial herbs; stems erect, often whitish, glabrous, to 1 m tall; basal leaves elliptic-lanceolate, acute at the apex, tapering to a winged petiole, to 10 cm long, to 4 cm wide, serrate to crenate, glabrous; cauline leaves alternate, linear-oblong to oblong to lanceolate to ovate, acute or obtuse at the apex, tapering to the sessile and more or less clasping base, serrate to crenate, glabrous, glandular-dotted, to 7 cm long, to 2 cm wide; heads several to numerous in a corymb, radiate, to 2.0 (–2.5) mm across; involucre hemispheric to globose to 10 mm high; phyllaries in 5 or 6 series, squarrose, reflexed to spreading to appressed, filiform to linear, usually strongly resinous; receptacle pitted, epaleate; ray flowers up to 35, rarely absent, pistillate, yellow, to 15 mm long, fertile; disc flowers numerous, tubular, yellow, bisexual, fertile or sometimes staminate; cypselae stramineous to grayish, thick, somewhat compressed, 4- or 5-sided, glabrous, 1.5–2.5 mm long; pappus of 2–8 deciduous awns or bristles shorter than the disc corollas.

Two varieties may be recognized in Illinois:

a. Upper and middle leaves 2–4 times longer than wide, ovate to oblong . . . . 1a. *G. squarrosa* var. *squarrosa*
a. Upper and middle leaves 5–8 times longer than wide, linear-oblong to oblanceolate . . . . 1b. *G. squarrosa* var. *serrulata*

1a. **Grindelia squarrosa** (Pursh) Dunal var. **squarrosa.** Fig. 66.
Upper and middle leaves 2–4 times longer than wide, ovate to oblong.

Common Name: Broad-leaved gum-plant; tarweed.
Habitat: Dry fields, disturbed soil.
Range: Native to the Great Plains and Rocky Mountains; adventive in many eastern states, Canada, and Asia.
Illinois Distribution: Occasionally in the northern half of Illinois, rare elsewhere.

This is the more common variety in Illinois, where it occurs in old fields and pastures, and along roadsides. It is introduced in Illinois.

This variety flowers from July to October.

66. *Grindelia squarrosa* (Broad-leaved gum-plant). a. Upper part of plant. b. Middle stem and leaves. c. Flowering head. d. Phyllary. e. Ray flower. f. Disc flower. g. Cypsela.

1b. **Grindelia squarrosa** (Pursh) Dunal var. **serrulata** (Rydb.) Steyerm. Ann. Mo. Bot. Gard. 21:227. 1934.

Grindelia serrulata Rydb. Bull. Torrey Club 31:646–647. 1904.

Upper and middle leaves 5–8 times longer than wide, linear-oblong to oblanceolate.

Common Name: Narrow-leaved squarrose gum-plant; tarweed.
Habitat: Dry disturbed soil.
Range: Native to the Great Plains and Rocky Mountains; adventive elsewhere, including Illinois.
Illinois Distribution: Scattered in the northern half of the state, rare in the southern half.

This variety flowers from July to October.

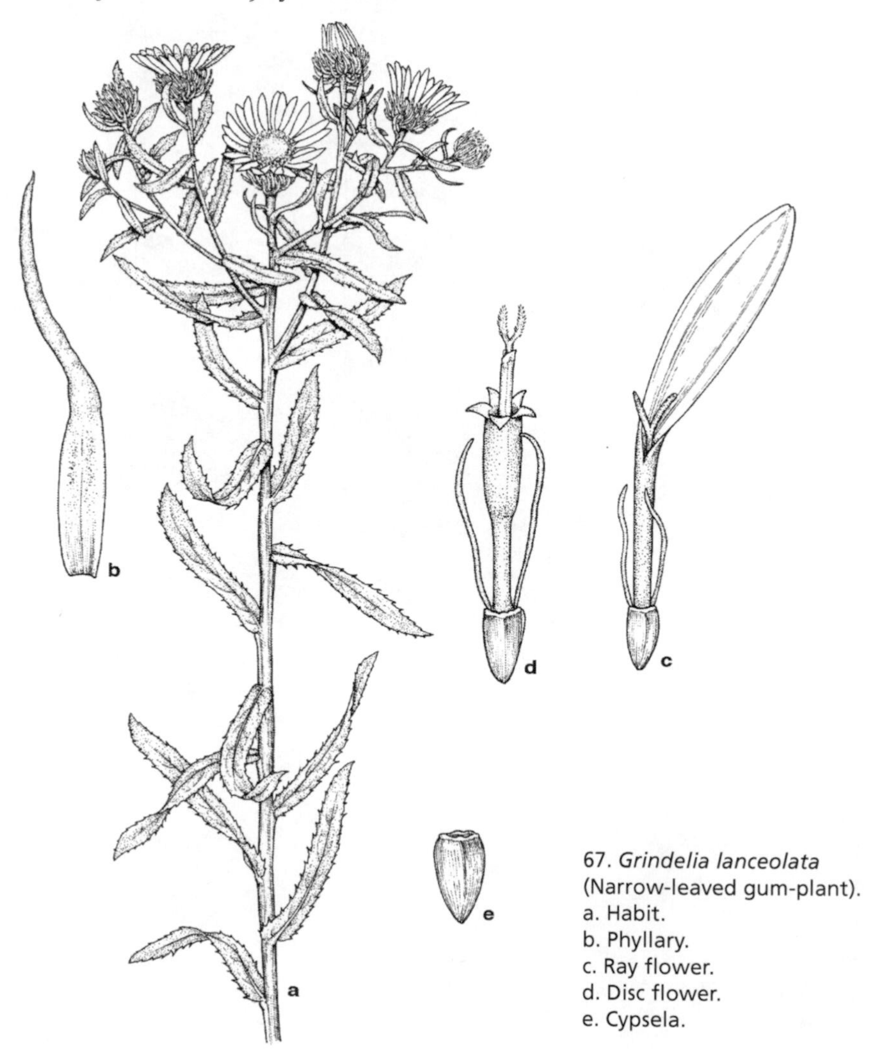

67. *Grindelia lanceolata* (Narrow-leaved gum-plant).
a. Habit.
b. Phyllary.
c. Ray flower.
d. Disc flower.
e. Cypsela.

2. **Grindelia lanceolata** Nutt. Journ. Acad. Nat. Sci. Phila. 7:73. 1834. Fig. 67.

Perennial herbs; stems erect, often whitish, glabrous except sometimes near the base, to 1.5 m tall; cauline leaves alternate, linear to lanceolate, acute to acuminate at the apex, tapering to the sessile and sometimes clasping base, glabrous, most or all of them spinulose-dentate, to 10 cm long, to 1.5 cm wide; heads several, in corymbs, radiate, up to 2.5 cm across; involucre more or less globose, 8–15 mm high; phyllaries usually in 5 or 6 series, slightly spreading to appressed, linear to narrowly lanceolate, usually somewhat resinous; receptacle pitted, epaleate; ray flowers up to 35, yellow, pistillate, 10–15 mm long; disc flowers numerous, yellow, bisexual, tubular; cypselae thick, stramineous or gray, 2–6 mm long, glabrous; pappus of 2 stiff awns 4–8 mm long, about as long as or longer than the disc corolla.

Common Name: Narrow-leaved gum-plant; spiny-toothed gum-plant.
Habitat: Sandy roadside (in Illinois).
Range: Illinois to Kansas, south to Texas and Alabama.
Illinois Distribution: Known only from Alexander County.

This species is similar to *G. squarrosa*, differing by its narrower leaves and its spreading to appressed phyllaries.

Grindelia lanceolata flowers from July to September.

3. **Grindelia perennis** A. Nels. Bull. Torrey Club 26:355–356. 1899. Fig. 68.
Grindelia squarrosa (Pursh) Dunal var. *quasiperennis* Lunell, Am. Midl. Nat. 3:143. 1913.

Perennial herbs; stems erect, glabrous or pubescent, resinous, to 30 cm tall; cauline leaves alternate, oblong to oblanceolate, obtuse to acute at the apex, tapering to the sessile or sometimes clasping base, entire or remotely serrulate, glandular-dotted, 3–6 cm long, 1–2 cm wide; heads in corymbs or panicles, radiate, up to 2 cm across; involucre globose to hemispheric, 8–15 mm high; phyllaries in 4 or 5 series, linear, at least the upper ones reflexed or spreading, squarrose, resinous at the tip; receptacle pitted, epaleate; ray flowers up to 60, yellow, pistillate, 10–20 mm long; disc flowers numerous, tubular, bisexual, yellow; cypselae white or gray, 4–6 mm long, glabrous; pappus of 2–4 subulate scales, shorter than the disc corolla.

Common Name: Perennial gum-plant.
Habitat: Disturbed sandy soil.
Range: Southwestern and western United States; adventive in Illinois.
Illinois Distribution: Known only from Alexander County.

This species is distinguished by its squarrose phyllaries and its merely serrate or sometimes entire leaves. Strother, in *Flora of North America*, has included this species and several others in a broad species known as *G. hirsutula* Hook. & Arn. (2006). There are enough differences in phyllary characters, pappus characters, leaf shape, and degree of resin to justify *G. perennis* as a distinct species.

Grindelia perennis flowers from July to September.

68. *Grindelia perennis* (Perennial gum-plant).

a. Upper part of plant.
b. Phyllary.
c. Ray flower.
d. Disc flower.
e. Cypsela.

19. **Symphyotrichum Nees**—Aster

Usually perennial herbs with rhizomes; stems 1 to several, erect to ascending to spreading, glabrous or pubescent; basal leaves sometimes present at flowering time; cauline leaves alternate, usually progressively smaller toward the top of the stem, short-petiolate to sessile to clasping, entire or serrate, glabrous or pubescent; heads few to several in panicles, radiate, the branches usually with several bracts; involucre cylindric to campanulate to hemispheric to turbinate; phyllaries in 3–9 series, usually unequal, appressed or less commonly squarrose, linear to oblanceolate to spatulate, glabrous or pubescent, usually green-tipped; receptacle flat, pitted, epaleate; ray flowers pistillate, white or purple or blue; disc flowers bisexual, yellowish at first, usually becoming pinkish or purplish with age, tubular; cypselae usually obovoid, usually flat, few- to several-nerved, glabrous or pubescent; pappus of barbellate capillary bristles not thickened at the tip and usually in 1 series.

Most of the species in Illinois that have been placed traditionally in *Aster* are now in *Symphyotrichum*. Our species have pappus with capillary bristles all alike in a single series.

This genus contains nearly 100 species, most of them in the United States.

Key to the Species of *Symphyotrichum* in Illinois

1. Basal or lower leaves cordate or subcordate, on long petioles.
  2. Cauline leaves clasping . . . . 19. *S. undulatum*
  2. Cauline leaves sessile or short-petiolate, not clasping.
    3. Leaves entire or nearly so.
      4. Leaves glabrous or nearly so above . . . . 18. *S. shortii*
      4. Leaves scabrous above.
        5. Phyllaries reflexed; leaves soft-hairy below . . . . 17. *S. anomalum*
        5. Phyllaries appressed; leaves rough-hairy below . . . . 20. *S. oolentangiense*
    3. Leaves, or most of them, regularly serrate.
      6. Petioles unwinged . . . . 12. *S. cordifolium*
      6. Petioles winged.
        7. Peduncles and branches of the inflorescence with few or no bracts; heads relatively few, fewer than 50; some or all the rays at least 10 mm long . . . . 16. *S. ciliolatum*
        7. Peduncles and branches of the inflorescence with many bracts; heads relatively many, usually more than 50; rays up to 8 (–10) mm long.
          8. Stems evenly and densely short-pilose or densely hirsute throughout; rays blue, lavender, or bluish white.
            9. Stems pilose; leaves firm, not becoming brittle with age; heads on short, ascending branches, the peduncles with a few scattered bracts; involucre campanulate, 4–7 mm high; ray flowers bright blue or lavender; cypselae glabrous . . . . 14. *S. drummondii*
            9. Stems densely hirsute; leaves membranous, becoming thicker and brittle with age; heads on long, spreading branches, the peduncles with many crowded bracts; involucre turbinate to hemispheric, 3.8–5.2 mm high; ray flowers bluish white; cypselae pubescent . . . . . . 15. *S. texanum*
          8. Stems glabrous or sparsely pubescent along decurrent lines on upper part; rays white . . . . 13. *S. urophyllum*

1. Basal or lower leaves neither cordate nor subcordate, sessile or petiolate.
  10. Stem leaves clasping or frequently auriculate at base.
    11. Leaves sericeous on both surfaces . . . . 9. *S. sericeum*
    11. Leaves glabrous or variously pubescent but not sericeous.
      12. Leaves coarsely serrate . . . . *S. prenanthoides*
      12. Leaves entire or finely and irregularly serrate.
        13. Stems glaucous.
          14. Leaves strongly clasping . . . . 21. *S. laeve*
          14. Leaves scarcely clasping . . . . 22. *S. concinnum*
        13. Stems green.
          15. Phyllaries glandular.
            16. At least the upper part of the stems strongly hirsute and often glandular.
              17. Leaves strongly auriculate-clasping . . . . . . 4. *S. novae-angliae*
              17. Leaves scarcely auriculate-clasping . . . . . . 3. *S. oblongifolium*
            16. Stems short-hairy, never glandular.
              18. Leaves thick and firm, up to 3½ times as long as wide.
                19. Involucre campanulate, 5.5–7.5 mm high; phyllaries in 4 or 5 series, squarrose, acute to acuminate at the apex, sparsely strigillose, densely stipitate-glandular, the middle phyllaries 1.0–1.2 mm wide. . . . . 6. *S. patens*
                19. Involucre turbinate, 8–12 mm high; phyllaries in 5–7 series, appressed, obtuse at the apex, densely strigillose, sparsely stipitate-glandular or eglandular, the middle phyllaries 1.2–1.7 mm wide . . . . 7. *S. patentissimum*
              18. Leaves thin and membranous, more than 3½ times longer than wide. . . . . 8. *S. phlogifolium*
          15. Phyllaries eglandular.
            20. None of the leaves more than 5 cm long; phyllaries uniformly pubescent. . . . . 5. *S. X amythistinum*
            20. Some or all the leaves at least 5 cm long; phyllaries glabrous or nearly so.
              21. Plants very slender, the stems at most 2.5 mm in diameter; phyllaries acute; rays white or pale lavender . . . . 31. *S. boreale*
              21. Plants stout, the stems more than 2.5 mm in diameter; phyllaries long-acuminate; rays blue (sometimes white in *S. firmum*).
                22. Stems coarsely hispid; rhizomes short . . . 32. *S. puniceum*
                22. Stems glabrous or nearly so; rhizomes elongated . . . . 33. *S. firmum*
  10. Leaves neither clasping nor auriculate at base.
    23. Annuals; plants along highways heavily salted in winter.
      24. Rays 4–7 mm long, 1.0–1.3 mm wide, in 1 series; disc flowers 25–50. . . . . 2. *S. divaricatum*
      24. Rays 1.5–4.0 mm long, 0.2–0.6 mm wide, in 1–3 series; disc flowers up to 20. . . . . 1. *S. subulatum*
    23. Perennials; plants usually not typically along highways heavily salted in winter.
      25. Leaves sericeous on both surfaces. . . . . 9. *S. sericeum*
      25. Leaves glabrous or pubescent but not sericeous.

26. Stems glandular, at least near the top; leaves oblong; phyllaries densely glandular . . . . 3. *S. oblongifolium*

26. Stems eglandular; leaves linear or lanceolate or elliptic, not oblong; phyllaries eglandular.
    27. Phyllaries uniformly pubescent on back.
        28. Heads numerous; involucre 3–5 mm high; rays 8–20 . . . . 10. *S. ericoides*
        28. Heads solitary to few; involucre 5–7 mm high; rays 20–35 . . . . 11. *S. falcatum*
    27. Phyllaries glabrous or at most ciliate, never uniformly pubescent on back.
        29. Phyllaries subulate-tipped, curved inward and more or less twisted at apex.
            30. Some or all the leaves 1 cm wide or wider; rays 16–35, 5–10 mm long . . . . 24. *S. pilosum*
            30. Leaves up to 6 mm wide; rays 12–16, 2–5 mm long . . . . 23. *S. parviceps*
        29. Phyllaries obtuse to acute, not subulate-tipped nor curved inward nor twisted at apex.
            31. Involucre up to 4 mm high.
                32. Bracteal leaves tiny, less than 1 cm long . . . . 26. *S. racemosum*
                32. Bracteal leaves larger, some of them at least 1.5 cm long . . . . 29. *S. lanceolatum*
            31. Involucre 4–12 mm high.
                33. Involucre 7–12 mm high, the phyllaries usually more than 1 mm wide (rarely less in *S. praealtum*); rays usually blue.
                    34. Rays 6–15; leaves conspicuously reticulate-veined below . . . . 30. *S. praealtum*
                    34. Rays 15–30; leaves not conspicuously reticulate-veined below.
                        35. Phyllaries obtuse; plants not glaucous; some of the involucres more than 9 mm high. . . . . 35. *S. turbinellum*
                        35. Phyllaries acute; plants more or less glaucous; none of the involucres more than 9 mm high . . . . 22. *S. concinnum*
                33. Involucre to 7 mm high, the phyllaries up to 1 mm wide; rays usually white (blue in *S. concinnum, S. praealtum,* and sometimes *S. boreale*).
                    36. Leaves conspicuously reticulate-veined below . . . . 30. *S. praealtum*
                    36. Leaves not conspicuously reticulate-veined below.
                        37. Leaves pubescent beneath, at least on the veins.
                            38. Leaves soft-hairy throughout on the lower surface . . . . 28. *S. ontarionis*
                            38. Leaves pubescent only on the midvein on the lower surface . . . . 27. *S. lateriflorum*
                        37. Leaves more or less glabrous.

39. Heads 1 to several, solitary at the tips of stiff, ascending peduncles. . . . . 25. *S. dumosum*

39. Heads few to many, on short peduncles.

40. Rays 30–50; most of the leaves less than 6 mm wide; heads few. . . . . 31. *S. boreale*

40. Rays usually fewer than 30; most of the leaves more than 6 mm wide; heads several to numerous.

41. Flowers blue; involucre 5–7 mm high . . . . 22. *S. concinnum*

41. Flowers usually white; involucre 3–4 mm high . . . . 29. *S. lanceolatum*

1. **Symphyotrichum subulatum** (Michx.) G. L. Nesom, Phytologia 77:293. 1995. Fig. 69.

Aster subulatus Michx. Fl. Bor. Am. 2:111. 1803.

Annual with a taproot and fibrous roots; stem 1, erect, to 75 cm tall, glabrous or nearly so; basal leaves absent at flowering time, oblanceolate to narrowly ovate, obtuse to acute at the apex, tapering to the sessile or petiolate base, entire to serrulate, glabrous or nearly so, usually ciliate, to 9 cm long, to 1.5 cm wide; cauline leaves linear to linear-lanceolate, acute at the apex, tapering to the sessile or short-petiolate base, entire or serrulate, glabrous or nearly so, up to 10 cm long, up to 1.0 (–1.5) mm wide; inflorescence an open panicle with 10–100 heads, the branches with several bracts; involucre 5–8 mm high, nearly as wide; phyllaries 20–30, in 3–5 series, unequal, subulate to linear to narrowly lanceolate, glabrous, greenish from tip to base and usually reddish or purple at the tip; receptacle flat, pitted, epaleate; ray flowers 15–30, 2–3 mm long (in Illinois), in 2 series, pink to light blue to white, pistillate; disc flowers up to 10, 3.8–5.0 mm long, bisexual, yellow; cypselae obovate to fusiform, flat, strigose, 5-nerved, 1.2–2.5 mm long, purple-brown; pappus of numerous white capillary bristles 3.5–5.5 mm long.

Common Name: Expressway aster (in Illinois).
Habitat: Adventive along highways (in Illinois).
Range: Native to the southeastern United States; adventive in Illinois and elsewhere.
Illinois Distribution: Becoming increasingly common in the northeastern counties; first collected in Illinois in 1982.

Although this is a species of the southeastern United States, particularly in brackish areas of the Coastal Plain, it has become fairly abundant along highways in the northeastern corner of Illinois where it occurs in areas that have been heavily salted during the winter months.

Symphyotrichum subulatum differs from *S. divaricatum* by its shorter and narrower rays and its fewer disc flowers.

In Illinois, it flowers from August to October.

69. *Symphyotrichum subulatum* (Expressway aster).
a. Upper part of plant.
b. Middle stem with leaf.
c. Flowering head.
d. Phyllary.
e. Ray flower.
f. Disc flower.
g. Cypsela.

2. **Symphyotrichum divaricatum** (Nutt.) G. L. Nesom, Phytologia 77:279. 1995. Fig. 70.

Tripolium divaricatum Nutt. Trans. Am. Phil. Soc. n.s. 7:296. 1840.

Aster subulatus Michx. var. *ligulatus* Shinners, Field & Lab. 21:159. 1953.

Symphyotrichum subulatum (Michx.) G. L. Nesom var. *ligulatum* (Shinners) S. D. Sundberg, Sida 21:907. 2004.

Annual with a taproot and fibrous roots; stem 1, erect, often branched above, to 1 m tall, glabrous or nearly so; basal leaves absent at flowering time, ovate to oblanceolate, obtuse to acute at the apex, tapering to the petiolate base, entire to serrulate, glabrous or nearly so, usually ciliate, to 7.5 cm long, to 1.2 cm wide; cauline leaves narrowly lanceolate, acuminate at the apex, tapering to the usually sessile base, entire or serrulate, glabrous or nearly so, up to 8 cm long, up to 2.5 mm wide; inflorescence an open panicle with spreading branches, with up to 100 secund heads, the branches with several bracts; involucre 5–7 mm high, nearly as wide;

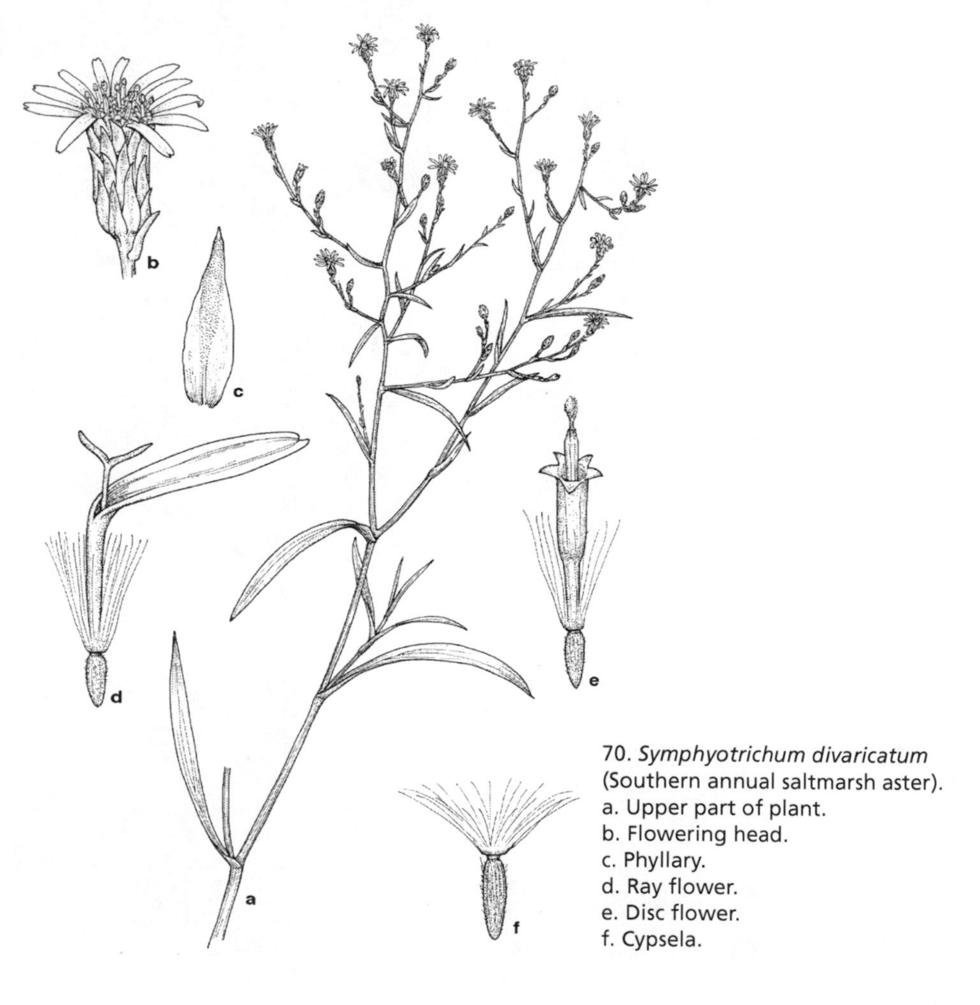

70. *Symphyotrichum divaricatum* (Southern annual saltmarsh aster).
a. Upper part of plant.
b. Flowering head.
c. Phyllary.
d. Ray flower.
e. Disc flower.
f. Cypsela.

phyllaries 25–45, in 3–5 series, unequal, subulate to lanceolate, glabrous, greenish at least near the tip; receptacle flat, pitted, epaleate; ray flowers 17–45, in 1 series, pistillate, 4–7 mm long, 1.0–1.3 mm wide, lavender or pale blue; disc flowers 25–50, 3.5–5.0 mm long, bisexual, yellow; cypselae obovate, flat, strigose, several-nerved, 1.5–3.0 mm long, usually dark purple; pappus of numerous white capillary bristles 3.0–4.5 mm long.

Common Name: Southern annual saltmarsh aster.
Habitat: Along roads (in Illinois).
Range: Tennessee to North Carolina, south to New Mexico, Texas, and Alabama; Mexico.
Illinois Distribution: Cook and DuPage counties.

This species is sometimes considered to be a variety of *S. subulatum*, but it differs by its lavender or pale blue ray flowers in a single series, its ray flowers 3.5–6.5 mm long and 1.0–1.3 mm wide, and 25–50 disc flowers.

Like *S. subulatum*, this species is usually associated with salt marshes in the southeastern United States, but its occurrence in Illinois is due no doubt to the abundant salt applied to roads in the northeastern part of the state during the winter.

Aster exilis may be the same as this species, but since the type specimen of *A. exilis* is lost, it is impossible to tell from the description if it is the same plant as *S. divaricatum*.

This species flowers in Illinois in August.

3. **Symphyotrichum oblongifolium** (Nutt.) G. L. Nesom, Phytologia 77:287. 1995. Fig. 71.
Aster oblongifolius Nutt. Gen. N. Am. Pl. 2:156. 1818.
Aster oblongifolius Nutt. f. *roseoligulatus* Shinners, Am. Midl. Nat. 26:417. 1941.
Aster oblongifolius Nutt. var. *angustatus* Shinners, Am. Midl. Nat. 26:418. 1941.

Perennial from creeping rhizomes and a short caudex; stems 1 to several, erect to ascending to spreading, much branched, to 1 m tall, evenly glandular-hirsute throughout; basal leaves absent at flowering time, sessile or nearly so, oblanceolate, entire, pubescent on both surfaces, often glandular, to 6 cm long, to 1.5 cm wide; cauline leaves numerous, progressively smaller, broadly linear to oblong, obtuse to acute at the apex, sessile or scarcely clasping at the base, entire, strigose or glabrous, to 4–10 cm long, 4–8 (–12) mm wide; inflorescence paniculate, with 10–25 heads 2–3 cm in diameter, the few to several bracts to 8 mm long; involucre 5–8 mm high, 5–10 mm wide; phyllaries in 4–6 series, subequal, linear, loosely spreading to reflexed at the tip, acute to acuminate, glandular, with long green tips; receptacle flat, pitted, epaleate; ray flowers 15–40, 7–15 mm long, pistillate, blue, rarely pink; disc flowers 30–50, 4.5–6.0 mm long, bisexual, reddish purple; cypselae oblong, flat, sericeous, 7- to 10-nerved, 1.5–2.5 mm long; pappus of numerous yellow-white to tan capillary bristles 4–6 mm long.

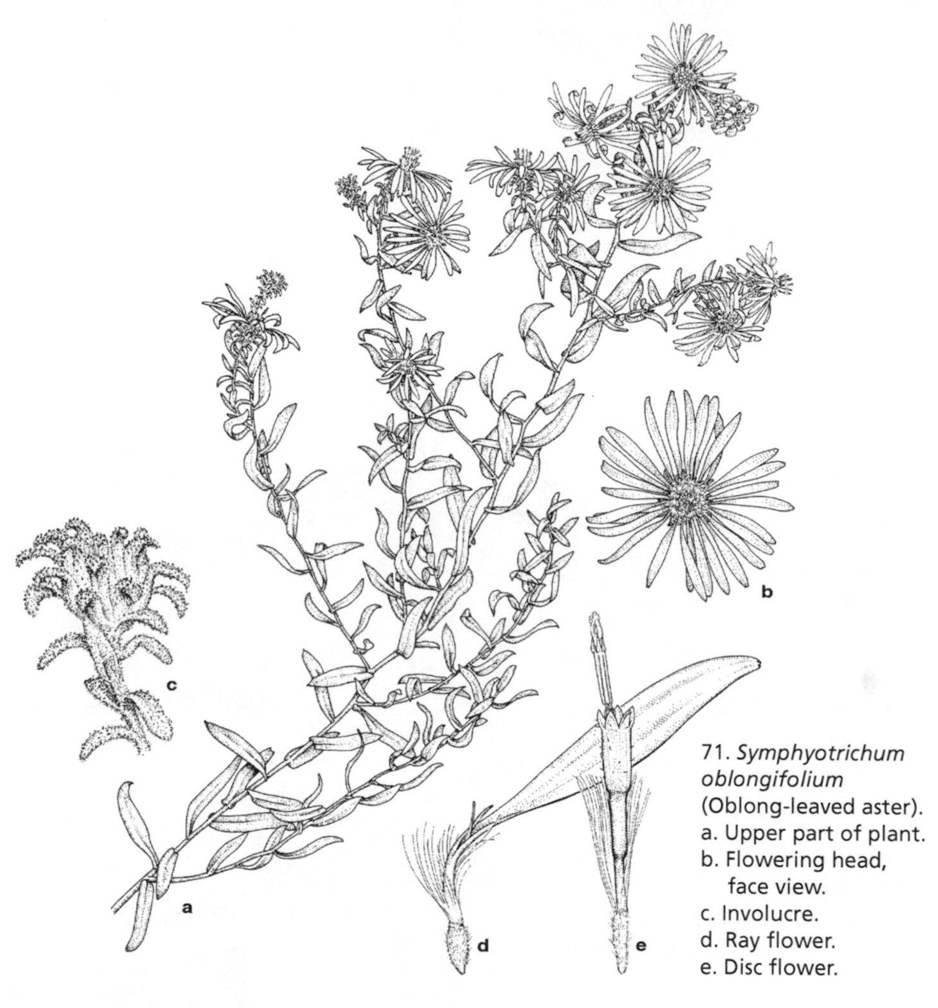

71. *Symphyotrichum oblongifolium* (Oblong-leaved aster).
a. Upper part of plant.
b. Flowering head, face view.
c. Involucre.
d. Ray flower.
e. Disc flower.

Common Name: Oblong-leaved aster; aromatic aster.
Habitat: Dry open woods, calcareous hill prairies, limestone barrens.
Range: Pennsylvania to Montana, south to New Mexico and Alabama.
Illinois Distribution: Uncommon to rare throughout most of the state, but apparently absent from the east-central counties.

This species is distinguished by its oblong, sessile or scarcely clasping leaves and its pubescent stems.

Plants with nearly linear leaves have been called var. *angustatus*. Pink-flowered plants are known as f. *roseoligulatus*. The transfers of these taxa have not been made to *Symphyotrichum* as yet.

This species resembles *S. X amethystinum*, but this latter plant has more pubescent leaves and stems and its phyllaries are not glandular.

This species flowers from August to October.

4. **Symphyotrichum novae-angliae** (L.) G. L. Nesom, Phytologia 77:287. 1995. Fig. 72.

Aster novae-angliae L. Sp. Pl. 2:875. 1753.
Aster roseus Desf. Tabl. Ecole Bot. ed. 3, 401. 1812.
Aster novae-angliae L. f. *roseus* (Desf.) Britt. Proc. Nat. Sci. Assoc. Staten Island 2. 1890.
Aster novae-angliae L. f. *geneseense* House, N.Y. State Mus. Bull. 243–244:40. 1923.

Perennial from a short, thickened rhizome and a hardened caudex; stems 1 to several, erect to ascending, up to 1.8 (–2.0) m tall, glandular-hispid to hirsute; basal leaves absent at flowering time, sessile or with short petioles, bracteolate, obtuse to subacute at the apex, tapering to the base, entire or sparsely serrate, pubescent on both surfaces, to 6 cm long, to 1.5 cm wide; cauline leaves numerous, progressively

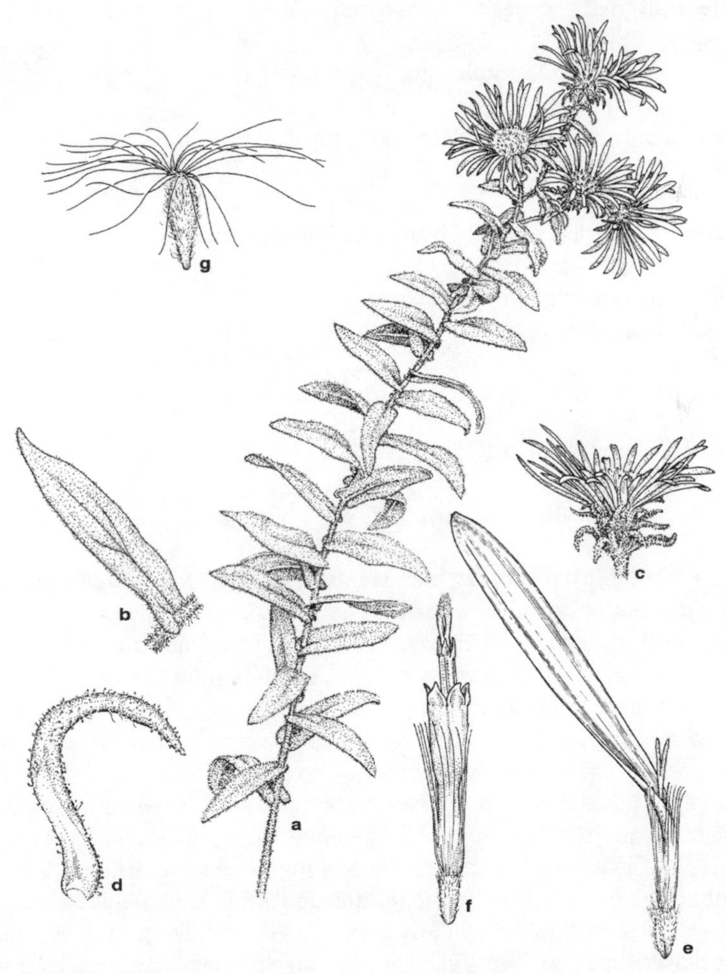

72. *Symphyotrichum novae-angliae* (New England aster).
a. Upper part of plant.
b. Section of stem with leaf.
c. Flowering head.
d. Phyllary.
e. Ray flower.
f. Disc flower.
g. Cypsela.

smaller, oblong-lanceolate to lanceolate, acute at the apex, strongly auriculate-clasping at the base, entire, strigose and scabrous above, softly strigose beneath, to 12 cm long, to 2 cm wide; inflorescence paniculate, with 30–65 heads, the heads 2.0–4.5 cm across, with a few bracts up to 12 mm long; involucre 6–10 mm high, 6–10 mm wide; phyllaries in (2–) 3–6 series, more or less equal, linear, glandular-hairy, often spreading or reflexed at the apex, the tips green or purplish; receptacle flat, pitted, epaleate; ray flowers 45–80 (–100), 10–25 mm long, pistillate, reddish purple, rarely rose or white; disc flowers 50–100, 4.5–6.5 mm long, bisexual, reddish purple; cypselae obovate, flat, sericeous, with 7–10 obscure nerves, 1.8–3.0 mm long; pappus of numerous tan to orange capillary bristles 4.0–6.5 mm long.

Common Name: New England aster.
Habitat: Mesic prairies, wet meadows, calcareous fens, pastures, along rivers and streams.
Range: New Brunswick to Manitoba and Washington, south to Oregon, Utah, New Mexico, Arkansas, and Georgia.
Illinois Distribution: Common in the northern two-thirds of Illinois, less common southward.

This species is readily distinguished by its clasping leaves with glandular-pubescent stems, leaves, and phyllaries.

Rose-flowered forms may be called f. *roseus*; white-flowered forms may be called f. *genesseensis*, although these have not been transferred to the genus *Symphyotrichum*.

Symphyotrichum novae-angliae flowers from late July to October.

5. **Symphyotrichum X amethystinum** (Nutt.) G. L. Nesom, Phytologia 77:275. 1995. Fig. 73.
Aster X amesthystinus Nutt. Trans. Am. Phil. Soc., n.s. 7:294. 1840.

Perennial from short, thick rhizomes; stems 1–5, erect, to 1.5 m tall, densely short-hairy to hirsute; lowest leaves usually early deciduous, pale green, oblanceolate, obtuse at the apex, sometimes mucronulate, tapering to the base, usually entire, to 40 mm long, 3–10 mm wide, strigose, sessile; cauline leaves pale green, linear to oblong, acute and often white-subulate at the apex, tapering to the sessile and partially clasping base, entire or rarely sparsely serrate, glabrous or sparsely strigose on both surfaces but harshly scabrous, to 50 mm long, to 3–10 mm wide; inflorescence paniculate, the branches ascending, with 30–80 sometimes secund heads; involucre turbinate, 4–7 mm high, 5–6 mm wide; phyllaries in 3–5 series, linear-lanceolate, hyaline on the margin, spreading or reflexed, the outer acute, puberulent and glandular, the inner acuminate, purplish, stipitate-glandular; receptacle flat, pitted, epaleate; ray flowers 15–35, 7–10 mm long, pistillate, blue or violet; disc flowers up to 30, bisexual, yellow to purplish; cypselae densely sericeous, several-nerved, 1.5–2.0 mm long; pappus of numerous tan capillary bristles 3.5–5.5 mm long.

Common Name: Amethyst aster.
Habitat: Moist ground.
Range: Vermont to Ontario and North Dakota, south to Nebraska, Missouri, Kentucky, and New Jersey; Colorado; Washington.
Illinois Distribution: Scattered in the northern half of the state.

This handsome aster is a hybrid between *S. ericoides* and *S. novae-angliae*. The involucres are intermediate in height between the smaller *S. ericoides* and the usually larger *S. novae-angliae*. The glandular outer phyllaries are like those of *S. ericoides*, while the glandular inner phyllaries are similar to those of *S. novae-angliae*. The rays are blue or violet as in *S. novae-angliae*.

Symphyotrichum X amethystinum flowers from June to October.

73. *Symphyotrichum X amethystinum* (Amethyst aster).
a. Upper part of plant.
b. Flowering head.
c. Involucre.
d. Phyllaries.
e. Ray flower.
f. Disc flower.
g. Cypsela.

6. **Symphyotrichum patens** (Ait.) G. L. Nesom, Phytologia 77:288. 1995. Fig. 74.
Aster patens Ait. Hort. Kew. 3:201. 1789.

Perennial from short rhizomes and a thickened caudex; stems 1 to a few, erect to ascending to spreading, to 1.0 (–1.5) m tall, densely short-pubescent; basal leaves absent at flowering time, oblanceolate, obtuse to acute at the apex, tapering to the sessile or short-petiolate base, entire or sparsely serrate, pubescent, scabrous, to 7 cm long, to 3 cm wide; cauline leaves progressively smaller, thick, firm, oblong to lance-ovate, acute at the apex, sessile and cordate-clasping at the base, entire, glabrous or pubescent and scabrous on the upper surface, pubescent on the lower surface, to 6 cm long, to 2 (–3) cm wide; inflorescence a panicle with single-headed branchlets, the branchlets with scalelike bracteal leaves up to 8 mm long, the heads 2.0–3.5 cm in diameter; involucre 5.5–7.5 mm high, 6–10 mm wide, campanulate;

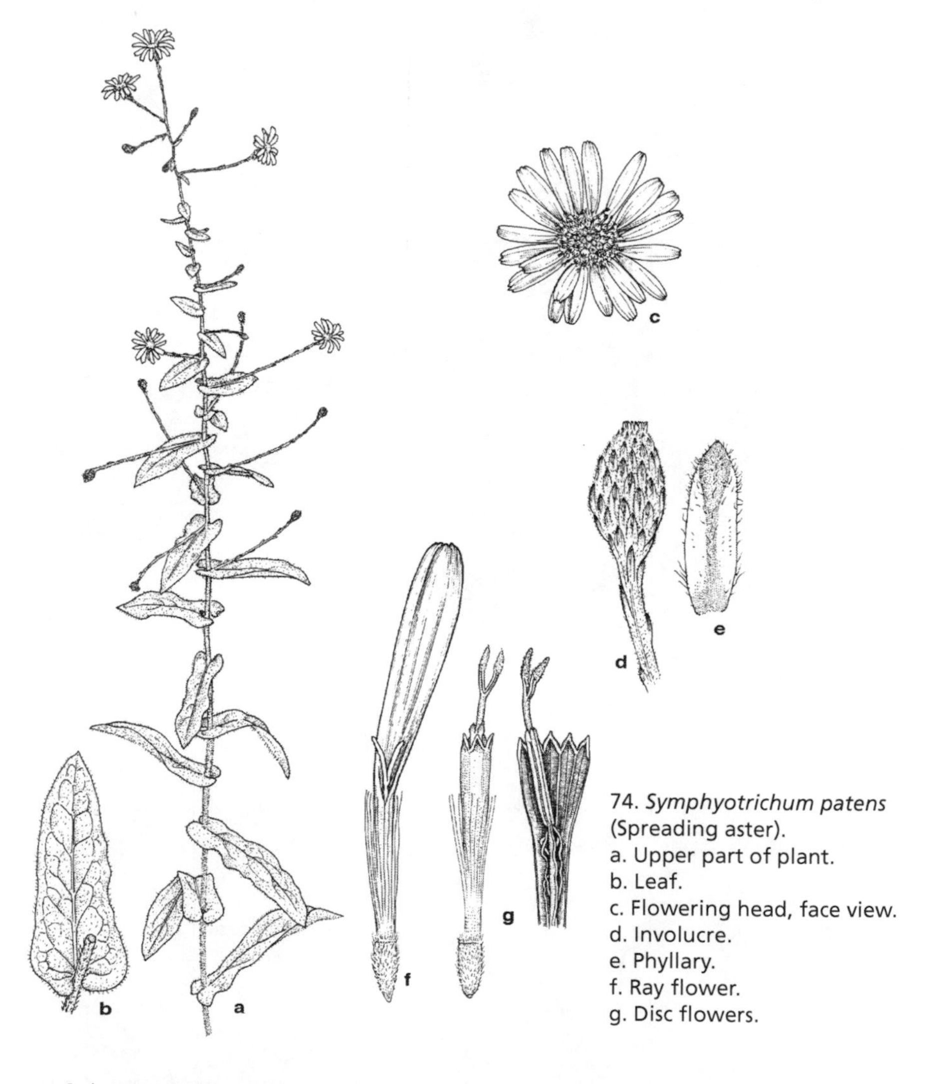

74. *Symphyotrichum patens* (Spreading aster).
a. Upper part of plant.
b. Leaf.
c. Flowering head, face view.
d. Involucre.
e. Phyllary.
f. Ray flower.
g. Disc flowers.

phyllaries in 4 or 5 series, unequal, squarrose, linear, acute to acuminate at the apex, sparsely strigillose, densely stipitate-glandular or eglandular, the middle phyllaries 1.0–1.2 mm wide; ray flowers 15–25, 8–15 mm long, pistillate, blue or purple; disc flowers 20–50, 3–4 mm long, bisexual, yellow; cypselae obovoid, not flat, sericeous, several-nerved, 2.0–3.5 mm long; pappus of numerous tawny capillary bristles 4.5–6.5 mm long.

Common Name: Spreading aster.
Habitat: Open woods.
Range: Maine to Illinois to Kansas, south to Texas and Florida.
Illinois Distribution: Occasional in the southern third of Illinois.

With its smaller campanulate involucres and its squarrose, densely glandular-stipitate phyllaries in 4 or 5 series, this species has several significant differences from *S. patentissimum*. *Symphyotrichum patens* is mostly a plant of the Appalachian, Piedmont, and Coastal Plain provinces of the southeastern United States, becoming less common in the western part of its range.

Symphyotrichum patens flowers in September and October.

7. **Symphyotrichum patentissimum** (Lindl. ex DC.) Mohlenbr. Vasc. Fl. Ill., ed. 4, 152. 2013. Fig. 75.
Aster patentissimus Lindl. ex DC. in A.P. DC. Prodr. 5:232. 1831.
Symphyotrichum patens (Ait.) G. L. Nesom var. *patentissimum* (Lindl. ex DC.) G. L. Nesom, Phytologia 77:285. 1995.

Perennial from short rhizomes and a thickened caudex; stems 1 to a few, erect to widely spreading, to 1.5 m tall, densely short-pubescent; basal leaves absent at flowering time, oblanceolate, obtuse to acute at the apex, tapering to the sessile or short-petiolate base, entire or sparsely serrate, pubescent, scabrous, to 7 cm long, to 3 cm wide; cauline leaves progressively smaller, thick, firm, oblong to lance-ovate, acute at the apex, sessile and cordate-clasping at the base, entire, glabrous or pubescent and scabrous on the upper surface, pubescent on the lower surface, to 6 cm long, to 3 cm wide; heads in a panicle with single-headed branchlets, the branchlets with scalelike bracteal leaves up to 10 mm long, the heads 2.0–3.5 cm in diameter; involucre 8–12 mm long, 8–12 mm wide, turbinate; phyllaries in 5–7 series, unequal, appressed, not squarrose, linear, obtuse at the apex, densely strigillose or sericeous, sparsely stipitate-glandular or eglandular, the middle phyllaries 1.2–1.7 mm wide; ray flowers 15–25, 8–15 mm long, pistillate, blue or purple; disc flowers 20–50, 3–4 mm long, bisexual, yellow, tubular; cypselae obovoid, not flat, sericeous, several-nerved, 2.0–3.5 mm long; pappus of numerous tawny capillary bristles 4.5–6.5 mm long.

Common Name: Western spreading aster.
Habitat: Dry woods.
Range: Kentucky to Kansas, south to Texas and Mississippi.
Illinois Distribution: Apparently confined to the southwestern counties of Illinois.

75. *Symphyotrichum patentissimum* (Western spreading aster).

a. Upper part of plant.
b. Flowering head, face view.
c. Involucre.
d. Phyllary.
e. Ray flower.
f. Disc flower.
g. Cypsela.

The large turbinate involucre and the appressed, scarcely glandular, densely strigillose or sericeous, obtuse phyllaries are very different from *S. patens*. It is generally more robust than *S. patens*.

Symphyotrichum patentissimum barely enters Illinois in the southwestern counties.

This plant flowers from August to October.

8. **Symphyotrichum phlogifolium** (Muhl. ex Willd.) G. L. Nesom, Phytologia 77:289. 1995. Fig. 76.

Aster phlogifolius Muhl. ex Willd. Sp. Pl. 3:2034. 1803.
Aster patens Ait. var. *phlogifolius* (Muhl. ex Willd.) Nees, Gen. Sp. Aster 49. 1832.

76. *Symphyotrichum phlogifolium* (Thin-leaved purple aster).
a. Upper part of plant.
b. Flowering head.
c. Phyllary.
d. Ray flower.
e. Disc flower.
f. Cypsela.

Cespitose perennial from short rhizomes and a thickened caudex; stem 1, erect to arching, to 1.0 (–1.2) m tall, softly pubescent with ascending to spreading hairs, occasionally stipitate-glandular; basal leaves absent at flowering time, obovate, acute at the apex, tapering to the sessile or very short-petiolate base, entire or sparsely serrate, scabrous, to 14 cm long, to 3 cm wide; cauline leaves thin, membranous, lanceolate to ovate, acute at the apex, sessile and strongly clasping at the base, entire, pubescent and scabrous on the upper surface, pubescent or sometimes stipitate-glandular on the lower surface, to 12 cm long, to 4 cm wide; inflorescence a panicle with single-headed branchlets, the branchlets with scalelike bracteal leaves; involucre 5–9 mm high, 5–10 mm wide; phyllaries in 4–6 series, unequal, linear, obtuse to acute, glandular-hairy, hyaline on the margins, erose, the green tips spreading; ray flowers 9–17, 8–18 mm long, pistillate, blue or purple; disc flowers about 15–35, 6–8 mm long, the lobes 1.0–1.5 mm long, bisexual, yellow; cypselae obovoid, not flat, sericeous, 7- to 10-nerved, 2.5–4.0 mm long; pappus of numerous tan or tawny capillary bristles 6–8 mm long.

Common Name: Thin-leaved purple aster.
Habitat: Woods.
Range: Massachusetts to Illinois, south to Alabama and Georgia.
Illinois Distribution: Known only from Hamilton County.

Although often considered to be a variety of *S. patens*, this plant is distinctive by its very thin, membranous leaves and its somewhat longer lobes of the disc corollas.

The Illinois station is apparently the westernmost location for this species in the central United States.

Symphyotrichum phlogifolium flowers in August and September.

9. **Symphyotrichum sericeum** (Vent.) G. L. Nesom, Phytologia 77:291. 1995. Fig. 77.
Aster sericeum Vent. Descr. Pl. Nouv. Pl. 33. 1800.

Cespitose perennial from a short rhizome and a thickened woody cormlike base; stems 1 to several, erect to ascending, to 1 m tall, densely canescent below, nearly glabrous above; basal leaves absent at flowering time, elliptic-lanceolate, acute at the apex, tapering to the sessile base, entire or sparsely serrate, pubescent on both surfaces, to 4 cm long, to 1.5 cm wide; cauline leaves alternate, oblong to narrowly lanceolate, obtuse to acute at the apex, rounded and sometimes slightly clasping at the base, entire, sericeous on both surfaces, to 3 cm long, to 8 mm wide, the upper ones progressively smaller; heads few to several in an open panicle with relatively few branches, radiate; involucre 5–10 mm high; phyllaries in 3–6 series, unequal, the outer ovate, the inner linear, usually scarious, spreading to squarrose, sericeous, usually green at the tip; receptacle flat, pitted, epaleate; ray flowers 10–30, 8–10 mm long, pistillate, purple to rose-purple; disc flowers up to 35, 5–7 mm high, bisexual, pink becoming purple; cypselae obovoid, not flat, 7- to 10-nerved, glabrous, 2–3 mm long, purple or brown; pappus of numerous tawny or white capillary bristles 6–7 mm long.

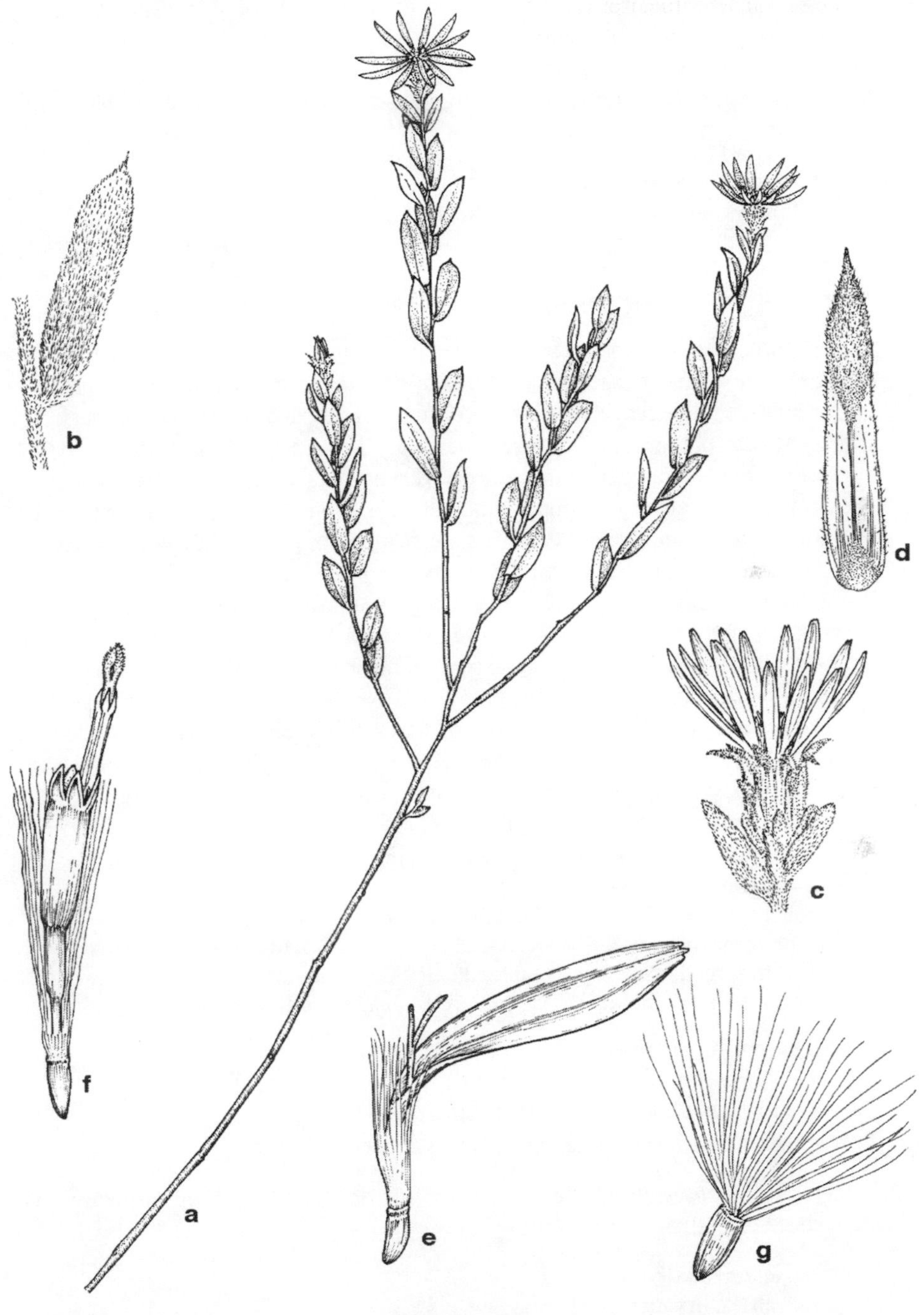

77. *Symphyotrichum sericeum* (Silky aster).
a. Upper part of plant.
b. Section of stem with leaf.
c. Flowering head.
d. Phyllary.
e. Ray flower.
f. Disc flower.
g. Cypsela.

Common Name: Silky aster.
Habitat: Dry prairies, sand barrens.
Range: Ontario to Manitoba, south to Texas and Georgia.
Illinois Distribution: Occasional in the northern half of Illinois, rare or absent elsewhere.

This species is readily distinguished by its leaves that are sericeous on both surfaces and by its sericeous phyllaries.

The flowers appear from August to October.

10. **Symphyotrichum ericoides** (L.) G. L. Nesom, Phytologia 77:280. 1995.
Aster ericoides L. Sp. Pl. 2:875. 1753.

Cespitose perennial from extensive rhizomes and stolons, usually in colonies; stems 1 to a few, erect to ascending, to 1 m tall, with appressed, ascending, spreading, or reflexed pubescence; basal leaves absent at flowering time, sessile or very short-petiolate, oblanceolate, obtuse to acute at the apex, tapering to the base, pubescent but becoming glabrate, entire or rarely sparsely serrate, to 6 cm long, to 1 cm wide, the middle and upper cauline leaves often absent at flowering time, linear to narrowly lanceolate, sessile, entire, to 5 cm long, to 8 mm wide, the uppermost leaves merging into subulate bracts; inflorescence paniculate, the branches loosely ascending, with numerous small heads 3–4 mm across, often secund on curving branches, the numerous bracts 2–8 mm long; involucre 3–5 mm high, 2–4 mm wide; phyllaries in 3 or 4 unequal series, imbricate, linear-lanceolate, densely pubescent to nearly glabrous on the back, ciliate on the margins, obtuse to acute but usually with a bristle tip, with a green, diamond-shaped apex; receptacle flat, pitted, epaleate; ray flowers 8–20, 3–6 mm long, pistillate, white or rarely bluish or pinkish; disc flowers up to 15, 2.5–4.0 mm long, bisexual, yellow or purple; cypselae oblong, flat, appressed-pubescent, purple or brown, faintly nerved, 1.0–2.2 mm long; pappus of numerous white capillary bristles 2–4 mm long.

Two varieties occur in Illinois.

a. Stems with appressed or ascending short hairs; phyllaries glabrous or nearly so or with appressed hairs on the back. . . . . 10a. *S. ericoides* var. *ericoides*
a. Stems with spreading or slightly reflexed hairs; phyllaries densely pubescent on the back with spreading hairs. . . . . 10b. *S. ericoides* var. *prostratum*

10a. **Symphyotrichum ericoides** (L.) G. L. Nesom var. **ericoides**. Fig. 78.
Aster multiflorus Ait. var. *caeruleus* Benke, Rhodora 30:78–79. 1928.
Aster ericoides L. f. *caerulea* (Benke) S. F. Blake, Rhodora 32:139. 1930.
Aster ericoides L. f. *gransii* Benke, Am. Midl. Nat. 13:328. 1932.

Stems with appressed or ascending short hairs; phyllaries glabrous or nearly so on the back or with appressed hairs.

Common Name: Heath aster.
Habitat: Prairies, dry disturbed areas.
Range: Quebec to British Columbia, south to Oregon, Arizona, Texas, Mississippi, and Virginia.
Illinois Distribution: Occasional to common throughout the state.

78. *Symphyotrichum ericoides* (Heath aster).
a. Upper part of plant.
b. Middle of stem with leaves.
c. Leaf.
d. Flowering head, side view.
e. Phyllaries.
f. Ray flower.
g. Disc flower.

This variety is distinguished by its usually uniformly pubescent phyllaries with a spine tip, its involucre 3–5 mm long, the rays 8–20 in number, and its numerous heads. *Symphyotrichum falcatum* is similar with its uniformly pubescent phyllaries, but its involucres are 5–7 mm long, its rays are 20–35 in number, and its heads are solitary or few.

Specimens with appressed or ascending pubescence on the stems are the typical variety, while specimens with spreading or reflexed pubescence on the stems may be called var. *prostratum*. Rarely blue-flowered specimens occur. They have been called f. *caeruleus*. The type specimen is from Cook County. Pink-flowered plants have been called f. *gransii*. Neither of these has been transferred to the genus *Symphyotrichum*. The type specimen is from Cook County.

Symphyotrichum ericoides var. *ericoides* flowers from July to October.

10b. **Symphyotrichum ericoides** (L.) G. L. Nesom var. **prostratum** (Kuntze) G. L. Nesom, Phytologia 77:281. 1995. Fig. 78.

Aster multiflorus Ait. var. *prostratus* Kuntze, Rev. Gen. Pl. 1:313. 1891.
Aster multiflorus Ait. var. *exiguus* Fern. Rhodora 1:187. 1899.
Aster exiguus (Fern.) Rydb. Bull. Torrey Club 28:505. 1901.
Aster ericoides (L.) var. *prostratus* (Kuntze) S. F. Blake, Rhodora 32:138. 1930.
Aster ericoides (L.) f. *prostratus* (Kuntze) Fern. Rhodora 51:46. 1949.

Stems with spreading or slightly reflexed hairs; phyllaries with spreading pubescence on the back.

Common Name: Heath aster.
Habitat: Prairies, dry disturbed areas.
Range: Quebec to British Columbia, south to Oregon, Arizona, Texas, Mississippi, and Virginia.
Illinois Distribution: Common throughout Illinois.

The spreading or slightly reflexed hairs on the stems differentiate this variety from the typical variety. This variety is common in Illinois.

Ahles in Jones & Fuller (1955) records a supposed hybrid between this variety and *S. pilosus* based on specimens from Champaign, Cook, and DuPage counties.

Symphyotrichum ericoides var. *prostratum* flowers from July to October.

11. **Symphyotrichum falcatum** (Lindl.) G. L. Nesom var. **commutatum** (Torr. & Gray) G. L. Nesom, Phytologia 77:281. 1995. Fig. 79.

Aster multiflorus Ait. var. *commutatus* Torr. & Gray, Fl. N. Am. 2:125. 1841.
Aster commutatus (Torr. & Gray) Gray, Syn. Fl. N. Am. 1:185. 1884.
Aster falcatus Lindl. var. *commutatus* (Torr. & Gray) A. G. Jones, Phytologia 62:131. 1987.
Symphyotrichum falcatum (Lindl.) G. L. Nesom ssp. *commutatum* (Torr, & Gray) Semple, Cult. Nat. Asters Ont. 134. 2002.

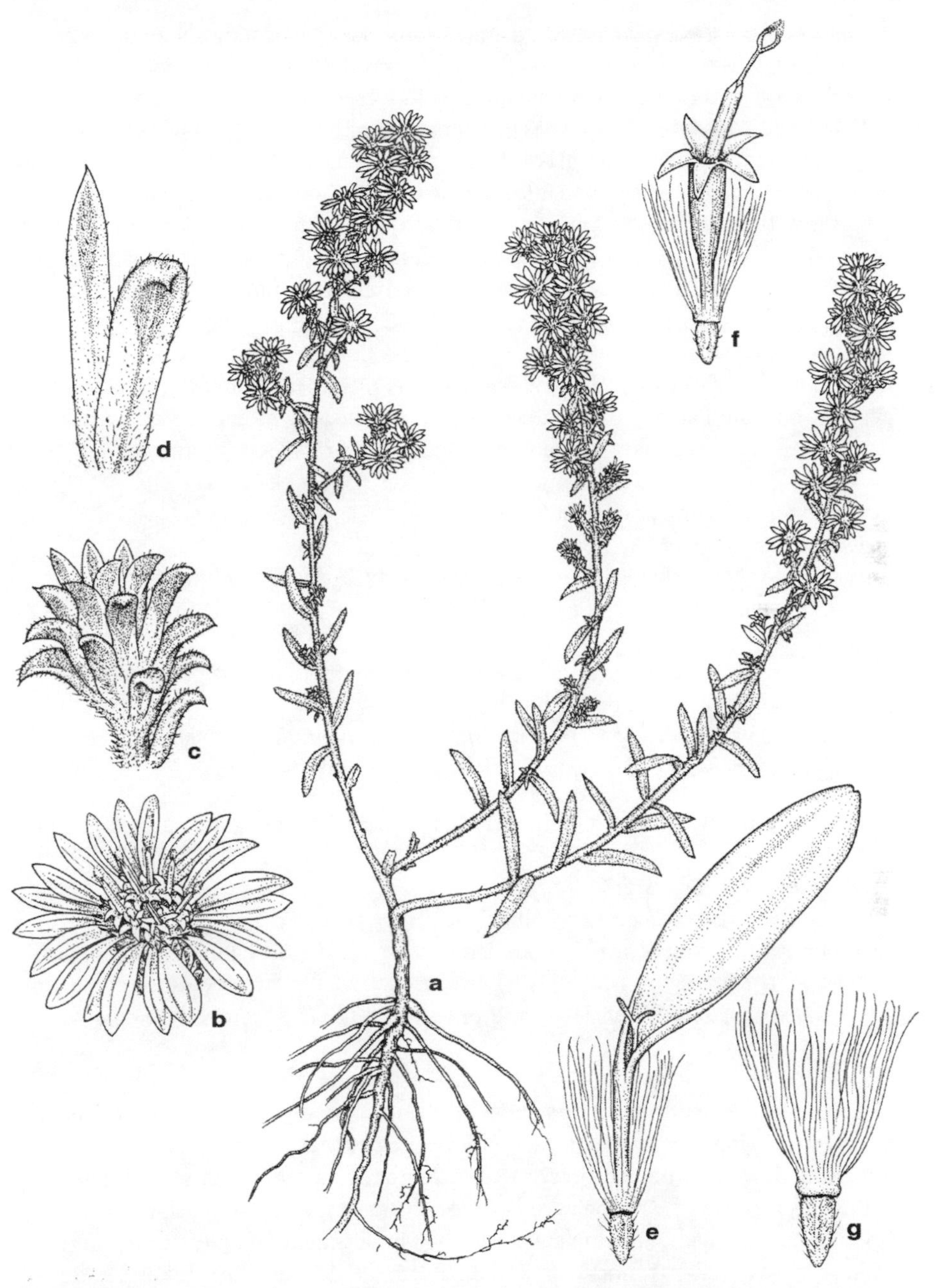

79. *Symphyotrichum falcatum* var. *commutatum* (White prairie aster).

a. Habit.
b. Flowering head, face view.
c. Involucre.
d. Phyllaries.
e. Ray flower.
f. Disc flower.
g. Cypsela.

Perennial from extensive rhizomes and stolons, usually forming colonies; stems solitary or few, erect to ascending, to 70 cm tall, densely pubescent except sometimes near the base; basal leaves absent at flowering time, sessile or on very short petioles, oblanceolate, obtuse to acute at the apex, tapering to the base, hispidulous but becoming glabrate, entire, 1–6 cm long, less than 1 cm wide; middle and upper cauline leaves progressively smaller, sessile, linear to narrowly oblong, obtuse to acute at the apex, tapering to the base, to 6 cm long, less than 1 cm wide, pubescent; inflorescence paniculate, the branches loosely ascending, the bracts linear, up to 8 mm long, with 15–45 heads, radiate; involucre hemispherical, 5–8 mm high, 6–8 mm wide; phyllaries in 3 or 4 series, linear, bristle-tipped, sometimes spreading, with a diamond-shaped green tip, puberulent; receptacle flat, pitted, epaleate; ray flowers 20–35, 4–9 mm long, pistillate, white or rarely pinkish; disc flowers up to 30, 3.5–5.0 mm long, bisexual, yellow; cypselae obovate, flat, appressed-puberulent, faintly nerved, brown or pale purple, 1.0–2.3 mm long; pappus of numerous white capillary bristles 3–5 mm long.

Common Name: White prairie aster.
Habitat: Prairies.
Range: Ontario to Saskatchewan, south to Arizona, Texas, and Illinois.
Illinois Distribution: Cook, DuPage, and Kane counties.

This plant is similar to *S. ericoides*, differing by larger involucres 5–7 mm long, more rays (20–35), and solitary or few heads.

The more western var. *falcatum* has short, nonrhizomatous rootstocks.

This taxon flowers from July to October.

12. **Symphyotrichum cordifolium** (L.) G. L. Nesom, Phytologia 77:278. 1995. Fig. 80.

Aster cordifolius L. Sp. Pl. 2:875. 1753.
Aster sagittifolius Wedemeyer ex Willd. Sp. Pl. 3:2033. 1803.
Aster cordifolius L. var. *sagittifolius* (Wedemeyer ex Willd.) A. G. Jones, Phytologia 63:131. 1987.
Symphyotrichum sagittifolium (Wedemeyer ex Willd.) G. L. Nesom, Phytologia 77:291. 1995.

Perennial from short rhizomes and a thickened caudex; stems 1 to several, erect, to 1.8 m tall, glabrous or with lines of pubescence beneath each leaf; leaves thin, present at flowering time, the lowest ovate, acute to acuminate at the apex, cordate at the base, coarsely serrate, glabrous and scabrous above, hirsute below, to 12 cm long, to 7 cm wide, on wingless or nearly wingless petioles up to 12 cm long; cauline leaves smaller, rounded or cordate at the sessile, nonclasping base; inflorescence an elongated, pyramidal panicle with 75–150 heads; involucre 3–6 mm high, 3–7 mm wide; phyllaries in several series, appressed, linear, obtuse at the apex, glabrous or sparsely short-hairy, with green or reddish tips; receptacle flat, pitted, epaleate; ray flowers 8–20, 5–12 mm long, pistillate, pale blue to purple; disc flowers up to 20, bisexual, cream to yellow, 3–5 mm long; cypselae obovate, flat, glabrous, usually light brown, 3- to 5-nerved, 1.5–2.5 mm long; pappus of numerous white capillary bristles 2.5–5.0 mm long.

Common Name: Heart-leaved aster.
Habitat: Dry woods.
Range: Nova Scotia to Minnesota, south to Missouri and Georgia.
Illinois Distribution: Occasional to common throughout Illinois.

80. *Symphyotrichum cordifolium* (Heart-leaved aster).
a. Upper part of plant.
b. Flowering head, face view.
c. Phyllary.
d. Ray flower.
e. Disc flower.
f. Cypsela.

This species is in the complex of species of *Symphyotrichum* that has thin, cordate, sharply serrate, nonclasping leaves and usually blue ray flowers. The others, recognized in this work from Illinois, are *S. ciliolatum, S. drummondii,* and *S. urophyllum*. Several botanists have included *S. sagittifolium* within *S. cordifolium* after Jones (1980) reported that she believes the type specimen of *Aster sagittifolius* belongs to *Aster cordifolius*. Plants heretofore called *A. sagittifolius* var. *dissitiflorus* E. S. Burgess in Illinois are now treated as *S. urophyllum*.

The absence of conspicuous winged petioles, the smaller involucres, and the generally obtuse phyllaries distinguish *S. cordifolium* from my concept of *S. urophyllum* and *S. drummondii.*

Symphyotrichum cordifolium flowers from August to October.

13. **Symphyotrichum urophyllum** (Lindl. ex DC.) G. L. Nesom, Phytologia 77:294. 1995. Fig. 81.

Aster urophyllus L. ex DC. in A.P. DC. Prodr. 5:233. 1836.

Aster sagittifolius Wedemeyer ex Willd. var. *dissitiflorus* E. S. Burgess, Ill. Fl. N. U. S. 3:365. 1898.

Perennial from short stout rhizomes and a thickened caudex; stems 1 to a few, erect to ascending, to 2 m tall, glabrous except for lines of pubescence beneath each leaf, or more or less pilose near the inflorescence; basal leaves present at flowering time, thin but not becoming brittle with age, ovate to lanceolate, acute at the apex, symmetrically cordate at the base, sharply serrate, glabrous or scabrous on the upper surface, pilose to villous on the lower surface, up to 15 cm long, up to 7 cm wide, on winged petioles; cauline leaves similar but becoming progressively smaller, the lower leaves ovate, acute at the apex, cordate at the base, on winged petioles, sharply serrate, glabrous or scabrous above, pilose below, to 7 cm long, to 3 cm wide, the upper ones narrowly ovate to lanceolate, sessile; inflorescence an elongated panicle with erect branches, with a few leaflike bracts to 1.2 cm long, with 75–150 heads 1.0–1.5 cm across, radiate; involucre cylindric, 4.0–6.5 mm high, about as wide; phyllaries in 3–6 series, unequal, narrowly lanceolate to oblong, acute, often spine-tipped, glabrous or sparsely pubescent, ciliate, with a narrow green tip; receptacle flat, pitted, epaleate; ray flowers 8–15 (–20), 4–10 mm long, pistillate, usually white; disc flowers up to 20 (–30), 3.5–5.0 mm long, bisexual, pale yellow or whitish; cypselae obovate, flat, glabrous or nearly so, 4- or 5-nerved, 1.5–2.5 mm long, purple-brown to brown; pappus of numerous white to cream capillary bristles 3.0–4.5 mm long.

Common Name: Arrow-leaved aster.
Habitat: Dry woods, mesic woods, prairies, pastures, old fields.
Range: Maine to North Dakota, south to Oklahoma, Missouri, and Illinois.
Illinois Distribution: Common throughout Illinois.

In previous works I have recognized *Aster urophyllus* and *A. sagittifolius* as distinct species. Since the only apparently valid difference between the two is the degree of pubescence on the stems, I am following others in combining the two in this publication. Since the type specimen of *A. sagittifolius* appears to be *A. cordifolius*, the epithet *urophyllum* is the correct one for this species.

Symphyotrichum urophyllum flowers from August to November.

81. *Symphyotrichum urophyllum* (Arrow-leaved aster).
a. Upper part of plant.
b. Flowering head.
c. Involucre.
d. Phyllary.
e. Ray flower.
f. Disc flower.
g. Cypsela.

14. **Symphyotrichum drummondii** (Lindl.) G. L. Nesom, Phytologia 77:279. 1995. Fig. 82.

Aster drummondii Lindl. Comp. Bot. Mag. 1:97. 1835.

Aster drummondii (Lindl.) G. L. Nesom var. *rhodactis* Benke, Rhodora 31:151. 1929.

Aster sagittifolius Wedem. ex Willd. var. *drummondii* (Lindl.) Shinners, Am. Midl. Nat. 26:406. 1941.

Perennial from short, stout rhizomes and a thickened caudex; stems 1 to a few, erect to ascending, to 1.2 m tall, densely pilose or sparsely hirsute, often nearly glabrous at base; lower leaves firm, not becoming brittle with age, broadly ovate to ovate-lanceolate, acute at the tip, cordate at the often asymmetrical base, serrate to dentate, usually scabrous above, pubescent beneath with spreading hairs, 6–15 cm

82. *Symphyotrichum drummondii* (Drummond's aster).

a. Upper part of plant.
b. Leaf.
c. Flowering head, face view.
d. Phyllary.
e. Ray flower.
f. Disc flower.

long, 3–6 cm wide, with narrowly winged petioles up to 6 cm long; cauline leaves similar but smaller, tapering to the base; inflorescence an open panicle with short ascending branches, the bracts few, not dense, leaflike, up to 12 mm long, with 50–150 heads; involucre campanulate, 4–7 mm high; phyllaries in 4 or 5 unequal series, more or less appressed, linear to narrowly lanceolate, acuminate, with a diamond-shaped green area at the tip, glabrous; receptacle flat, pitted, epaleate; ray flowers 10–20, 7–10 (–12) mm long, pistillate, bright blue to lavender; disc flowers up to 20, 3.5–5.0 mm high, bisexual, yellowish; cypselae obovate, flat, glabrous, pale brown, weakly nerved, 2–3 mm long; pappus of numerous white or tan or cream capillary bristles 3–4 mm long.

Common Name: Drummond's aster.
Habitat: Moist forests, dry open woods, prairies, fields, pastures, glades, edge of cliffs.
Range: Pennsylvania to Minnesota and Nebraska, south to Texas and Georgia.
Illinois Distribution: Common throughout the state.

This species is similar to *S. urophyllum* and *S. texanum* because of its cordate leaves and winged petioles. It differs from *S. urophyllum* by its evenly and densely pilose stems, its often asymmetrical cordate leaf bases, and its campanulate involucres. The flowers of *S. urophyllum* are usually white. It is also similar to *S. texanum*, but *S. texanum* has symmetrical cordate leaf bases, densely hirsute stems, long-spreading peduncles with densely crowded bracts, turbinate to hemispheric involucres, and strigillose cypselae. The flowers of *S. texanum* are bluish white.

The flowers bloom from August to late October.

15. **Symphyotrichum texanum** (E. S. Burgess) Semple, Univ. Waterloo Biol. Ser. 41: 134. 2002. Fig. 83.

Aster texanum E. S. Burgess in Small, Fl. S.E.U.S. 1214. 1903

Symphyotrichum drummondii (Lindl.) G. L. Nesom var. *texanum* (E. S. Burgess) G. L. Nesom, Phytologia 77:279. 1999.

Perennial herb from short, stout rhizomes and a thickened caudex; stems 1 to a few, erect, to 80 cm tall, densely hirsute; lower leaves membranous, becoming thicker and brittle with age, broadly ovate to ovate-lanceolate, acute at the tip, cordate at the symmetrical base, serrate to dentate, usually scabrous above, pubescent beneath, to 15 cm long, to 6 cm wide, with broadly winged petioles up to 6 cm long; cauline leaves alternate, similar but smaller, sometimes tapering to the base; inflorescence an open panicle with long, spreading branches, the peduncles with densely crowded bracts up to 12 mm long, with up to 120 heads; involucre turbinate to hemispheric, 3.8–5.2 mm high; phyllaries in 4 or 5 unequal series, more or less appressed, linear to narrowly lanceolate, acuminate at the apex, with diamond-shaped green areas near the tip; receptacle flat, pitted, epaleate; ray flowers 10–15, 7–10 (–12) mm long, pistillate, pale bluish white; disc flowers up to 20, 3.5–5.0 mm high, bisexual, yellowish; cypselae obovate, flat, strigillose, brown, usually nerved, 2–3 mm long; pappus of numerous white or tan capillary bristles 3–4 mm long.

83. *Symphyotrichum texanum* (Texas aster).

a. Upper part of plant.
b. Flowering head.
c. Phyllary. d. Ray flower.
e. Disc flower.
f. Cypsela.

Common Name: Texas aster.
Habitat: Crevices of limestone cliff (in Illinois).
Range: Illinois to Kansas, south to Texas and Mississippi.
Illinois Distribution: Known only from Monroe County.

Sometimes considered to be a variety of *S. drummondii*, this is, I believe, a distinct species. It differs in several significant ways from *S. drummondii*. Its stems are densely hirsute, its cordate leaf bases are symmetrical, it has long-spreading panicle branches with many crowded bracts on the peduncles, its involucres are turbinate or hemispheric, its flowers are pale bluish white, and its cypselae are strigillose.

I have crawled over numerous limestone cliffs that tower above the Mississippi River many times, often encountering *S. drummondii*. When James Ozment and I came upon a specimen in a limestone crevice in Monroe County, we knew it was different from *S. drummondii* because it had pale bluish white flower heads and densely hirsute stems. On further observations, we found that this plant had a shorter stature, long-spreading branches of the inflorescence with many crowded bracts on the peduncles, thin membranous basal leaves, and strigillose cypselae. We determined it to be *Aster texanus* (at the time; later transferred to *Symphyotrichum*).

Symphyotrichum texanum flowers from August to October.

16. **Symphyotrichum ciliolatum** (Lindl.) A. Love & D. Love, Taxon 31:359. 1982. Fig. 84.
Aster ciliolatus Lindl. in W. J. Hook. Fl. Bor. Am. 2:9. 1854.

Perennial from long rhizomes or a thickened caudex; stems 1 to a few, erect, to 1.2 m tall, glabrous in the lower half, hirsute in the upper half and in the inflorescence; leaves thin, the basal absent at flowering time, ovate, acute to acuminate at the tip, cordate or abruptly contracted at the base, glabrous on the upper surface, pubescent on the lower surface, serrate, to 15 cm long, 2–6 (–7) cm wide, the winged petiole 3–12 cm long and often clasping; upper leaves gradually reduced, lanceolate to linear, entire, sessile and often clasping; inflorescence paniculate, the branches ascending, with few or no bracts, with 3–50 heads; involucre hemispheric, 3–6 (–7) mm high, 6–8 mm wide, the phyllaries in 4 or 5 series, linear to narrowly oblong, with a green diamond-shaped acuminate tip and often a green midvein, glabrous but with ciliate margins; receptacle flat, pitted, epaleate; ray flowers 10–20, 7–15 mm long, pistillate, bluish; disc flowers up to 25, 4.5–6.5 mm long, bisexual, reddish purple; cypselae flat, obovate, glabrous or sparsely pubescent, pale yellow, 5- to 6-nerved, 1–2 mm long; pappus of white to pinkish plumose bristles 3–6 mm long.

Common Name: Fringed aster.
Habitat: Woods.
Range: Nova Scotia to Yukon, south to Wyoming, northern Illinois, and New York.
Illinois Distribution: Known only from Cook and DuPage counties.

With its heart-shaped basal leaves and winged petioles, this species is similar to *S. drummondii*, *S. texanum*, and *S. urophyllum*. It differs by having few or no bracts on the branches of the inflorescence, its flowering heads usually fewer than 50, and its ray flowers 10–20 mm long.

The flowers bloom in August and September.

84. *Symphyotrichum ciliolatum* (Fringed aster).

a. Upper part of plant.
b. Lower leaf.
c. Flowering head.
d. Phyllaries.
e. Ray flower.
f. Disc flower.
g. Cypsela.

17. **Symphyotrichum anomalum** (Engelm. ex Torr. & Gray), G. L. Nesom, Phytologia 77:275. 1995. Fig. 85.

Aster anomalus Engelm. ex Torr. & Gray, Fl. N. Am. 2:503. 1843.

Aster anomalus Engelm. ex Torr. & Gray f. *albidus* Steyerm. Rhodora 51:119. 1949.

Perennial from short, stout rhizomes; stems 1 to a few, erect to ascending, to 1 m tall, pubescent and scabrous, or nearly glabrous near the base; lower leaves present at flowering time, ovate to lanceolate, acute at the tip, cordate at the base, glabrous or scabrous on the upper surface, soft-hairy or less commonly rough-hairy on the lower surface, entire or nearly so, to 9 cm long, to 5 cm wide, the petiole usually winged, to 8 cm long; upper leaves smaller, lanceolate, sessile or subsessile, the petioles, if present, winged, acute at the tip, tapering or occasionally cordate at the

85. *Symphyotrichum anomalum* (Manyray aster).

a. Upper part of plant.
b. Middle stem with leaves.
c. Flowering head.
d. Ray flower.
e. Disc flower.

base, often with several reduced leaves in axillary tufts; inflorescence open, paniculate, the branches loosely ascending, the bracts leaflike, to 2.5 cm long, narrowly lanceolate, with 10–70 heads; involucre hemisperical, 5–10 mm high, 5–15 mm wide, the heads usually long-stalked, radiate; phyllaries in several overlapping series, loosely imbricate, densely pubescent, the green tips spreading or reflexed, with a narrow diamond-shaped green tip; receptacle flat, pitted, epaleate; ray flowers up to 45, 5–18 mm long, pistillate, bright blue or lavender; disc flowers up to 45, bisexual, purple, 4.0–5.5 mm long; cypselae glabrous, flat, purple-brown, 3- or 5-ribbed, 1–3 mm long; pappus of plumose bristles 3.5–5.0 mm long, stramineous to tan.

Common Name: Manyray aster.
Habitat: Prairies, dry woods.
Range: Illinois to Kansas, south to Oklahoma and Arkansas.
Illinois Distribution: Southwestern Illinois north to Woodford and Peoria counties.

This species is distinguished by its cordate leaves with winged petioles, the phyllaries spreading or reflexed, and blue ray flowers up to 45 in number.

Specimens with white rays occur occasionally and may be called f. *albiflora*. This forma has not been transferred as yet to *Symphyotrichum*.

The flowers bloom in September and October.

18. **Symphyotrichum shortii** (Lindl.) G. L. Nesom, Phytologia 77:291. 1995. Fig. 86.
Aster shortii Lindl. in Hook. Fl. Bor. Am. 2:9. 1834.
Aster shortii Lindl. f. *gronemannii* Benke, Rhodora 31:150. 1929.

Cespitose perennial from a short rhizome and a thickened woody caudex; stems 1 to a few, erect to ascending, up to 1.2 m tall, short-hirsute, becoming glabrate; basal leaves absent at flowering time, lance-ovate to ovate, obtuse to acute at the apex, cordate at the base, crenate-serrate, glabrous and often scabrous on the upper surface, glabrous to short-hirtellous on the lower surface, to 6 cm long, to 3.5 cm wide; cauline leaves narrowly ovate to lanceolate, acuminate at the apex, cordate at the base, entire or sparsely serrate, glabrous and often scabrous on the upper surface, glabrous to short-hirtellous on the lower surface, to 10 cm long, to 6 cm wide, on wingless petioles; inflorescence an open panicle with spreading branches having many leafy bracts and 70–140 heads, radiate; involucre 4–6 mm high, nearly as wide; phyllaries in 4–6 series, unequal, linear to lanceolate, obtuse to acute, strigillose, scarious, erose, with a diamond-shaped green tip; receptacle flat, pitted, epaleate; ray flowers 10–20, 8–15 mm long, pistillate, blue, rarely rose; disc flowers up to 30, 4.5–7.0 mm long, bisexual, yellow, usually becoming purple; cypselae obovate, flat, glabrous, 3- to 7- nerved, 2.5–3.5 mm long, purple or brown; pappus of numerous tawny capillary bristles 4–6 mm long.

Common Name: Short's aster.
Habitat: Rocky woods, mesic woods.

Range: New York and Ontario to Minnesota, south to Arkansas, Mississippi, and Florida.
Illinois Distribution: Occasional to common throughout the state.

This species is distinguished by its cordate lower leaves on long petioles and its entire leaves that are glabrous on the upper surface.

Rose-flowered plants are rarely seen. They may be called f. *gronemannii*, but this forma has not been transferred to *Symphyotrichum*.

Symphyotrichum anomalum is a similar species, but it has leaves that are scabrous above and soft-hairy beneath and phyllaries that are spreading to reflexed.

This species flowers from August to October.

86. *Symphyotrichum shortii* (Short's aster).

a. Upper part of plant.
b. Middle stem with leaves.
c. Flowering head.
d. Phyllary.
e. Ray flower.
f. Disc flower.
g. Cypsela.
h. Involucre.

19. **Symphyotrichum undulatum** (L.) G. L. Nesom, Phytologia 77:293. 1995. Fig. 87.
Aster undulatus L. Sp. Pl. 2:875. 1753.

Cespitose perennial from short rhizomes and a thickened woody caudex; stems 1 to a few, erect, to 1.0 (–1.5) m tall; basal leaves absent at flowering time, broadly lanceolate to ovate, acute to acuminate at the apex, cordate or subcordate at the base, entire or serrate, glabrous and scabrous on the upper surface, villous on the lower surface, up to 15 cm long, up to 7 cm wide, on wingless petioles up to 10 cm long; cauline leaves progressively smaller, often entire, sessile and clasping at the base; inflorescence an open panicle with 40–100 radiate heads, with several appressed, subulate bracts 2–3 mm long; involucre 3.5–5.5 mm high, nearly as wide; phyllaries in 4–6 series, unequal, appressed to spreading, linear to oblong-lanceolate, obtuse to acute to acuminate, puberulent, ciliate, scarious, erose, green-tipped; receptacle flat, pitted, epaleate; ray flowers 8–20, 5–12 mm long, pistillate, blue or purple; disc flowers up to 25, 4–6 mm long, bisexual, yellow, becoming purple; cypselae oblong, flat, strigillose, 3- or 4-nerved, 1.7–2.2 mm long, brown to purple; pappus of numerous cream or orange capillary bristles 2.5–5.0 mm long.

Common Name: Wavyleaf aster.
Habitat: Dry woods.
Range: Nova Scotia to Ontario, south to Illinois, Louisiana, and Florida.
Illinois Distribution: Scattered in the southeastern counties.

This is the only species of *Symphyotrichum* in Illinois with cordate and long-petiolate basal leaves and clasping cauline leaves.

This species flowers from August to October.

20. **Symphyotrichum oolentangiense** (Riddell) G. L. Nesom, Phytologia 77:288. 1995.
Aster oolentangiensis Riddell, Med. Phys. Sci. 8:495. 1835.
Aster azureus Lindl. Comp. Bot. Mag. 1:98. 1835.
Aster azureus Lindl. f. *laevicaulis* Fern. Rhodora 51:94. 1949.

Perennial from rhizomes and a hardened caudex; stems 1 to a few, erect to ascending, to 1.5 m tall, glabrous to puberulent to hispid; basal leaves present at flowering time, ovate to broadly lanceolate, acute to acuminate at the apex, cordate or subcordate at the base, entire or sometimes minutely serrate, hispid and scabrous on the upper surface, pilose on the lower surface, to 18 cm long, to 7 cm wide, the more or less winged petiole to 10 cm long; cauline leaves progressively smaller, lanceolate to narrowly ovate, acute at the apex, cordate or rounded at the base, entire or nearly so, glabrous, to 6 cm long, to 2 cm wide; inflorescence open and paniculate, with long spreading branches, the bracts to 12 mm long, with 20–150 heads 1.5–2.5 cm across, radiate; involucre 4.5–8.0 mm high, 5–10 mm wide; phyllaries in 3–6 series, unequal, linear to narrowly lanceolate with appressed green tips, obtuse to acute at the apex, ciliate on the margins; receptacle flat, pitted, epaleate; ray flowers 10–25, 6–12 mm long, pistillate, blue, rarely white; disc flowers 15–30, 4–5 mm long, bisexual, yellow; cypselae obovoid, flat, glabrous or nearly so, 1.2–2.0 mm long, purple-brown, weakly 3- to 5-nerved; pappus of numerous white or tan capillary bristles 3.5–5.0 mm long.

Two varieties occur in Illinois:

a. Stems scabrous and puberulent . . . . 20a. *S. oolentangiense* var. *oolentangiense*

a. Stems glabrous or nearly so . . . . 20b. *S. oolentangiense* var. *laevicaule*

87. *Symphyotrichum undulatum* (Wavyleaf aster).

a. Upper part of plant.
b. Middle part of stem.
c. Flowering head.
d. Involucre.
e. Phyllary.
f. Ray flower.
g. Disc flower.
h. Cypsela.

20a. **Symphyotrichum oolentangiense** (Riddell) G. L. Nesom var. **oolentangiense.** Fig. 88.

Stems scabrous and puberulent.

Common Name: Azure aster.
Habitat: Prairies, glades, black oak savannas.
Range: New York to Ontario to Minnesota and South Dakota, south to Texas and Florida.
Illinois Distribution: Occasional throughout the state.

88. *Symphyotrichum oolentangiense* (Azure aster).

a. Upper part of plant.
b. Lower part of stem.
c. Flowering head.
d. Involucre.
e. Phyllary.
f. Ray flower.
g. Disc flower.
h. Cypsela.

This species for years was known as *Aster azureus*, but Riddell's epithet *oolentangiensis* barely predates Lindley's *azureus*.

This species differs by its sessile, scabrous, usually entire cauline leaves, its appressed phyllaries, and its azure blue rays. Typical var. *oolentangiense* has stems that are scabrous and puberulent. White-flowered plants rarely occur.

This variety flowers from August to November.

20b. **Symphyotrichum oolentangiense** (Riddell) G. L. Nesom var. **laevicaule** (Fern.) Mohlenbr. Guide Vasc. Fl. Ill., ed. 4, 152. 2013. Fig. 88.

Aster azureus Lindl. f. *laevicaulis* Fern. Rhodora 51:94. 1949.

Aster oolentangiense Riddell var. *laevicaulis* (Fern.) A. G. Jones, Bull. Torrey Club 110:41. 1983.

Common Name: Smooth-stemmed azure aster.

Habitat: Prairies, woods.

Range: New York to South Dakota, south to Texas and Florida.

Illinois Distribution: Not common in the southern half of Illinois.

This variety, which sometimes occurs with the typical variety, has stems that are glabrous or nearly so. It flowers from August to November. The type specimen for f. *laevicaulis* is from Winnebago County.

21. **Symphyotrichum laeve** (L.) A. Love & D. Love, Taxon 31:359. 1982. Fig. 89.

Aster laevis L. Sp. Pl. 2:876. 1753.

Aster cyaneus Hoffm. Phyt. Blatt. 1:71. 1803.

Aster laevis L. f. *beckwithiae* House, N.Y. State Mus. Bull. 254. 704. 1924.

Perennial from short, stout to long-creeping rhizomes and a thickened caudex; stems 1 to a few, erect to ascending, to 1.2 m tall, glabrous except sometimes for lines of pubescence below each leaf, glaucous, sometimes reddish near the base; basal and lowest cauline leaves absent at flowering time, oblanceolate to narrowly ovate, obtuse to acute at the apex, tapering to the base, sessile or short-petiolate, entire or nearly so, to 20 cm long, to 3 cm wide; middle and upper cauline leaves progressively smaller, broadly ovate to linear, acute at the apex, sessile and strongly clasping at the base, thick, firm, entire or shallowly serrate, glabrous, to 15 cm long, to 1 cm wide, the uppermost bractlike, clasping; inflorescence an open panicle with ascending branches, with 5–50 heads, radiate, the bracts linear to narrowly lanceolate, to 8 mm long; involucre hemispheric, 5–9 mm high, 6–9 mm wide; phyllaries in several series, unequal, linear to narrowly oblong, obtuse to acute to acuminate, glabrous except for the margin, with short, diamond-shaped green tips; receptacle flat, pitted, epaleate; ray flowers 15–25 (–30), 8–15 mm long, pistillate, blue or purple, rarely white; disc flowers up to 50, bisexual, yellow, 4.5–6.5 mm long; cypselae obovate, flat, glabrous or minutely pubescent, with 3–5 nerves, 1.3–2.5 mm long, purple-brown; pappus of numerous, often reddish, capillary bristles 4.5–6.0 mm long.

Common Name: Smooth aster.
Habitat: Prairies, glades, rock ledges, occasionally in fens; roadsides.
Range: New Brunswick to Manitoba, south to Kansas, Arkansas, and Georgia.
Illinois Distribution: Common throughout Illinois.

The distinguishing characteristics of this species are its strongly clasping cauline leaves, its glabrous and glaucous stems, and its entire to sparsely serrate leaves.

Rare white-flowered forms have been called f. *beckwithiae*, but this forma has not been transferred to *Symphyotrichum*.

Jones & Fuller (1955) report a specimen from McHenry County to be a hybrid between this species and *S. lanceolatum*.

Symphyotrichum laeve flowers from July to October.

89. *Symphyotrichum laeve* (Smooth aster).
a. Upper part of plant.
b. Node with leaf.
c. Flowering head.
d. Phyllaries.
e. Ray flower.
f. Disc flower.

22. **Symphyotrichum concinnum** (Willd.) Mohlenbr. Guide Vasc. Fl. Ill., ed. 4, 150. 2013. Fig. 90.
Aster concinnus Willd. Enum. Pl. 2:884. 1809.
Symphyotrichum laeve (L.) G. L. Nesom var. *concinnum* (Willd.) G. L. Nesom, Phytologia 77: 283. 1995.

Perennial with short, stout rhizomes and a thickened caudex; stems erect, to 1 m tall, glabrous or sometimes with lines of pubescence below each leaf, not glaucous; lowest leaves broadly lanceolate with winged petioles, early deciduous; cauline leaves thick, firm, pale green, linear to lanceolate, acute at the tip, sessile or rarely weakly clasping at the base, usually entire or occasionally serrate, glabrous,

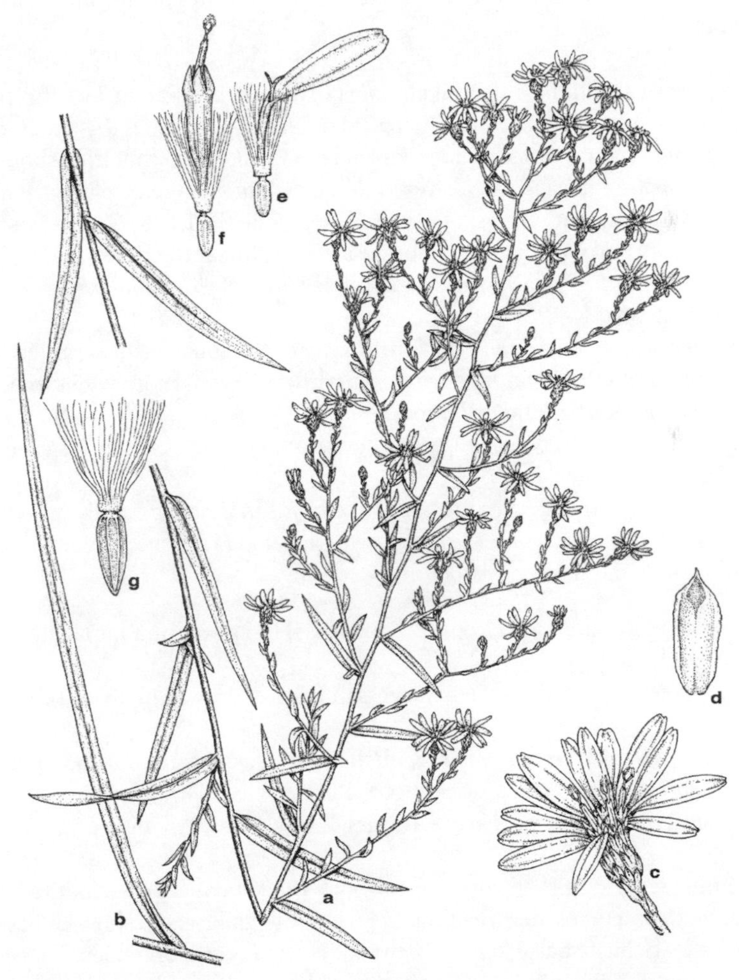

90. *Symphyotrichum concinnum* (Narrow-leaved smooth aster). a. Upper part of plant. b. Node with leaf. c. Flowering head. d. Phyllary. e. Ray flower. f. Disc flower. g. Cypsela.

the uppermost bractlike; inflorescence an open panicle with 10–50 heads, radiate; involucre hemispherical, 5–7 mm high, 4–7 mm wide; phyllaries in several series, linear, obtuse to acute to acuminate, green-tipped, glabrous; receptacle flat, pitted, epaleate; ray flowers 15–25, 10–15 mm long, pistillate, blue or purple; disc flowers about 50, bisexual, yellow; cypselae obovate, flat, glabrous or puberulent, 1.3–1.5 mm long; pappus of numerous, sometimes reddish, capillary bristles 3–5 mm long.

Common Name: Narrow-leaved smooth aster.
Habitat: Rich woods.
Range: Pennsylvania and District of Columbia to Kentucky and southeastern Illinois, south to Mississippi and Florida.
Illinois Distribution: Known from the southeastern counties of Hardin, Pope, and Saline.

Although several botanists, as well as *Flora of North America*, consider this plant to be a variety of *S. laeve*, the morphological differences between the two and the seemingly habitat and range differences lead me to conclude that they should be considered distinct species. The leaves of *S. concinnum* are linear to linear-lanceolate and are barely or not at all clasping at the base, while the leaves of *S. laeve* are lanceolate to ovate and conspicuously auriculate-clasping at the base. In addition, the stems and leaves of *S. laeve* are glaucous, while those of *S. concinnum* are scarcely or not at all glaucous.

Symphyotrichum concinnum occurs primarily in the southeastern United States, particularly in mountain and mesic woods, while *S. laeve* is more widespread and grows mostly in prairies and dry woods.

Symphyotrichum concinnum begins to flower in September, while *S. laeve* often flowers in Illinois in mid-July.

The location for this species in Hardin County is in a rich, cove-like woods where it occurs with *Silene ovata* and *Micranthes virginiana*, two other Appalachian species that are known in Illinois only from this location.

23. **Symphyotrichum parviceps** (E. S. Burgess) G. L. Nesom, Phytologia 77:288. 1995. Fig. 91.

Aster ericoides L. var. *parviceps* E. S. Burgess in Britt. & Brown, Ill. Fl. N. U. S. 3:379. 1898.
Aster parviceps (E. S. Burgess) Mack. & Bush, Man. Fl. Jackson Co. 196. 1902.
Aster depauperatus Fern. var. *parviceps* (E. S. Burgess) Fern. Rhodora 10:94. 1908.
Aster pilosus Willd. ssp. *parviceps* (E. S. Burgess) A. G. Jones, Phytologia 55:381. 1983.

Perennial with a short rhizome and a short, thickened, woody caudex; stems 1 to a few, erect to ascending to spreading, to 80 cm tall, glabrous to short-pubescent to hirsute, often reddish; basal leaves absent at flowering time, oblanceolate, acute at the apex, rounded or tapering to the sessile or short-petiolate base, entire, glabrous or sparsely pubescent, to 4 cm long, to 8 mm wide; cauline leaves progressively smaller, linear-lanceolate to linear, acute at the apex, tapering to the sessile base, entire or sometimes serrate, to 8 cm long, to 6 mm wide; inflorescence a panicle with loosely spreading branches, with linear bracts up to 8 mm long, with 50–100 often secund heads 7–12 mm in diameter, radiate; involucre 3–5 mm high, 3–4 mm wide; phyllaries in 3–5 series, unequal, glabrous, linear to narrowly lanceolate, with in-rolled, green, subulate, twisted tips; receptacle flat, pitted, epaleate; ray flowers 10–18, 3–5 mm long, pistillate, white; disc flowers up to 12 (–15), 2–3 mm long, bisexual, pale violet to yellow; cypselae obovate, flat, 3- to 4-nerved, sparsely pubescent, gray or tan, 0.8–1.5 mm long; pappus of numerous white capillary bristles 2–3 mm long.

91. *Symphyotrichum parviceps* (Small white aster).
a. Upper part of plant.
b. Leaf.
c. Margin of leaf with teeth.
d. Flowering head.
e, f. Phyllaries.
g. Ray flower.
h. Disc flower.

Common Name: Small white aster.
Habitat: Prairies, open woods, roadsides.
Range: Illinois to Iowa, south to Oklahoma and Arkansas.
Illinois Distribution: Occasional in the northern three-fifths of the state.

This species and *S. pilosum* are similar in that they both have phyllaries that are subulate with an in-rolled twisted apex. Both have white ray flowers. *Symphyotrichum parviceps* differs by its narrower leaves and fewer and shorter rays. It apparently intergrades with *S. pilosum* on occasion.

Symphyotrichum parviceps has one of the most restricted ranges of any species of the genus in Illinois.

The flowers appear from August to October.

24. **Symphyotrichum pilosum** (Willd.) G. L. Nesom, Phytologia 77:289. 1995.
Aster pilosus Willd. Sp. Pl. 3:2025. 1803.

Cespitose perennial from long-creeping rhizomes and a thickened caudex; stems 1 to several, stout, erect to ascending, to 1.0 (–1.5) m tall, hispid or hirsute or occasionally glabrous; basal leaves absent at flowering time, oblanceolate to obovate, obtuse to acute at the apex, tapering to the winged petiolate base, pilose or less commonly glabrous, crenate-serrate, up to 6 (–10) cm long, to 1.5 (–2.0) cm wide; cauline leaves progressively smaller, narrowly elliptic to linear to subulate and often aristate at the apex, tapering to the sessile base, entire or nearly so, pilose or less commonly glabrous, up to 10 cm long, up to 1.5 cm wide; inflorescence an open panicle with ascending to spreading branches, with 50 or more often secund heads, radiate, the bracts linear to subulate; involucre 3–6 mm high, 3–5 mm wide; phyllaries in 4 or 5 series, unequal, appressed to spreading, linear to subulate, glabrous, scarious, erose, the green tips marginally in-rolled and twisted; receptacle flat, pitted, epaleate; ray flowers (10–) 15–35, 5–10 mm long, pistillate, white; disc flowers up to 60, 2.5–5.5 mm long, bisexual, yellow; cypselae oblong, flat, pubescent, 4- to 6- nerved, gray, 1.0–1.5 mm long; pappus of numerous white capillary bristles 3.5–4.0 mm long.

Two varieties occur in Illinois:

a. Stems and leaves pubescent; branches of the inflorescence rather long .... 24a. *S. pilosum* var. *pilosum*
a. Stems and leaves glabrous; branches of the inflorescence short. . . . 24b. *S. pilosum* var. *pringlei*

24a. **Symphyotrichum pilosum** (Willd.) G. L. Nesom var. **pilosum.** Fig. 92.
Aster villosus Michx. Fl. Bor. Am. 2:113. 1803, *non* Thunb. (1800).
Aster ericoides L. var. *villosus* (Michx.) Torr. & Gray, Fl. N. Am. 2:124. 1841.
Aster ericoides L. var. *platyphyllus* Torr. & Gray, Fl. N. Am. 2:124. 1841.
Aster pilosus Willd. var. *platyphyllus* Blake, Rhodora 32:139. 1930.
Aster pilosus Willd. var. *demotus* Blake, Rhodora 32:139. 1930.

Stems and leaves pubescent; branches of the inflorescence long.

Common Name: Hairy white aster.
Habitat: Dry or wet soil, old fields, prairies, roadsides, woods.
Range: New Brunswick to Minnesota and South Dakota, south to Oklahoma, Missouri, and Florida; British Columbia.
Illinois Distribution: Common throughout the state.

92. *Symphyotrichum pilosum* var. *pilosum* (Hairy white aster).
a. Habit.
b. Node with leaf.
c. Flowering branch.
d. Involucre.
e. Phyllary.
f. Ray flower.
g. Disc flower.

Because of the subulate, in-rolled, twisted tips of the phyllaries, this species is related to *S. parviceps*. It differs from *S. parviceps* by its wider leaves and longer, more numerous rays.

Until 1943, Illinois botanists called this plant *Aster ericoides*.

Plants with cauline leaves 1.5–2.0 cm wide have been called var. *platyphyllus*, but this variety has not been transferred to *Symphyotrichum*.

This is one of the most common plants of *Symphyotrichum* in Illinois, where it grows in a wide variety of habitats. It flowers from August to November.

92 (*continued*). *Symphyotrichum pilosum* var. *pringlei* (Pringle's aster).

h. Upper part of plant.
i. Basal rosette with leaves.
j. Leaf.
k. Flowering head.
l. Phyllary.
m. Ray flower.
n. Disc flower.

24b. **Symphyotrichum pilosum** (Willd.) G. L. Nesom var. **pringlei** (Gray) G. L. Nesom, Phytologia 77:289. 1995. Fig. 92.
Aster pringlei Gray, Britt. Ill. Fl. N. U.S. 3:379. 1858.
Aster ericoides L. var. *pringlei* Gray, Syn. Fl. N. Am. 1:184–185. 1884.

Stems and leaves glabrous; branches of the inflorescence short.

Common Name: Pringle's aster.
Habitat: Dry sandy soil.
Range: Nova Scotia to Ontario to Minnesota, south to Illinois, Kentucky, and North Carolina.
Illinois Distribution: Known only from Lake County.

This variety looks very different from var. *pilosum* because of its shorter inflorescence branches and its glabrous stems and leaves. However, there is little difference in the structures of its flower heads from var. *pilosum*. It flowers from August to October.

25. **Symphyotrichum dumosum** (L.) G. L. Nesom, Phytologia 77:280. 1995. Fig. 93.
Aster dumosus L. Sp. Pl. 873. 1753.
Aster coridifolius Michx. ex Willd. Sp. Pl. 3:2028. 1804.
Aster fragilis Lindl. Bot. Reg. 18:pl. 1537. 1832.
Aster dumosus L. var. *strictior* Torr. & Gray, Fl. N. Am. 2:128. 1841.
Aster dumosus L. var. *coridifolius* (Michx. ex Willd.) Torr. & Gray, Fl. N. Am. 2:128. 1841.
Symphyotrichum dumosum (L.) G. L. Nesom var. *strictior* (Torr. & Gray) G. L. Nesom, Phytologia 77:280. 1995.

Perennial from long, slender, creeping rhizomes and a thickened caudex; stems usually solitary, erect to ascending, to 1 m tall, usually much branched above, puberulent with curled hairs, or glabrous; lowest leaves early deciduous, oblanceolate, obtuse to acute at the apex, tapering to the base, glabrous or slightly scabrous, to 5 cm long, to 1.5 cm wide, short-petiolate; cauline leaves narrowly elliptic to oblanceolate to linear, obtuse but often mucronate at the apex, tapering to the sessile base, entire or sparsely serrate, scabrous on the upper surface otherwise glabrous, to 10 cm long, to 8 mm wide, those on the branches of the inflorescence numerous and often 3–4 mm long and up to 1 mm wide; inflorescence paniculate, with ascending branches, with heads in small clusters or solitary, radiate; involucre 4–7 mm high, 3–6 mm wide; phyllaries in 4 or 5 series, oblong to linear-lanceolate, acute at the apex, glabrous or minutely puberulent, with a narrow diamond-shaped green tip less than half the length of the phyllary; receptacle flat, pitted, epaleate; ray flowers 9–30, 4–9 mm long, pistillate, white or pale lavender; disc flowers 15–30, 3.5–4.5 mm long, bisexual, yellow or brownish; achenes obovate, flat, strigose, few-nerved, 1.2–2.5 mm long; pappus of numerous white capillary bristles 3.5–4.5 mm long.

Common Name: Bushy aster; rice button aster.
Habitat: Moist woods, calcareous fens, sand flats, prairies, along roads.
Range: Maine to Ontario to Wisconsin, south to Iowa, Texas, and Florida.
Illinois Distribution: Locally common and scattered in Illinois.

This species is somewhat variable, and two of the variations have been found in Illinois. Plants with the inflorescence occupying a tenth to half the height of the plant and with nearly glabrous stems have been called var. *strictior*, while plants with the inflorescence occupying one-third to two-thirds the height of the plant and pubescent stems are known as var. *dumosum*. Variety *coridifolius* apparently is the same as var. *dumosum*.

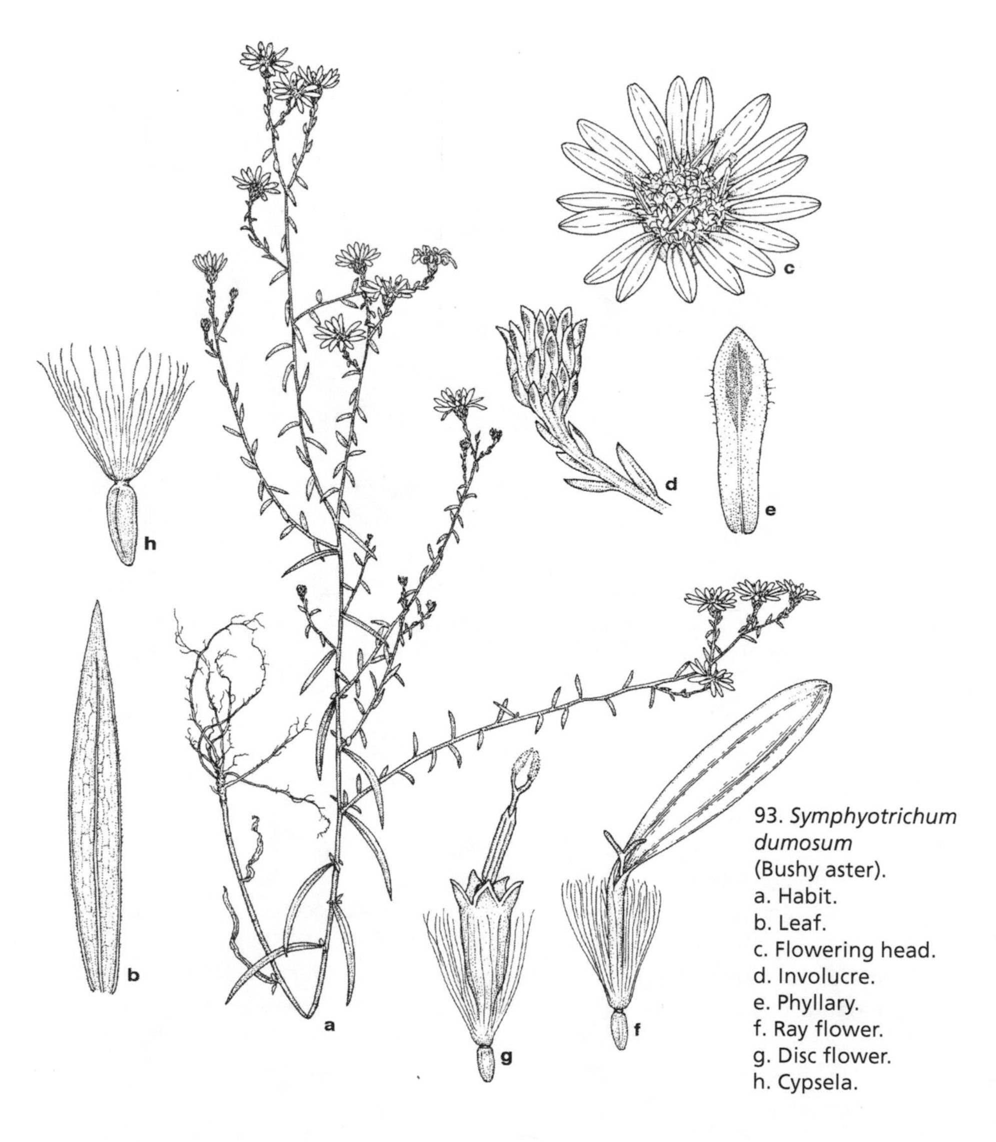

93. *Symphyotrichum dumosum* (Bushy aster).
a. Habit.
b. Leaf.
c. Flowering head.
d. Involucre.
e. Phyllary.
f. Ray flower.
g. Disc flower.
h. Cypsela.

Symphyotrichum dumosum is distinguished by its numerous small leaves on the branches of the inflorescence, its involucre 4–7 mm high, its phyllaries less than 1 mm wide, and its usually glabrous leaves. It also resembles *S. racemosum* var. *subdumosum* because of its numerous heads and tiny bracts on the inflorescence branches, but the green tips of the phyllaries in *S. dumosum* are less than half the length of the phyllaries.

This species flowers from August to October.

26. **Symphyotrichum racemosum** (Ell.) G. L. Nesom var. **subdumosum** (Wieg.) G. L. Nesom, Phytologia 77:291. 1995. Fig. 94.

Aster racemosus Ell. Sketch Bot. S. Car. 2:348–349. 1823.

Aster vinimeus Lam. var. *subdumosus* Wieg. Rhodora 30:171. 1928.

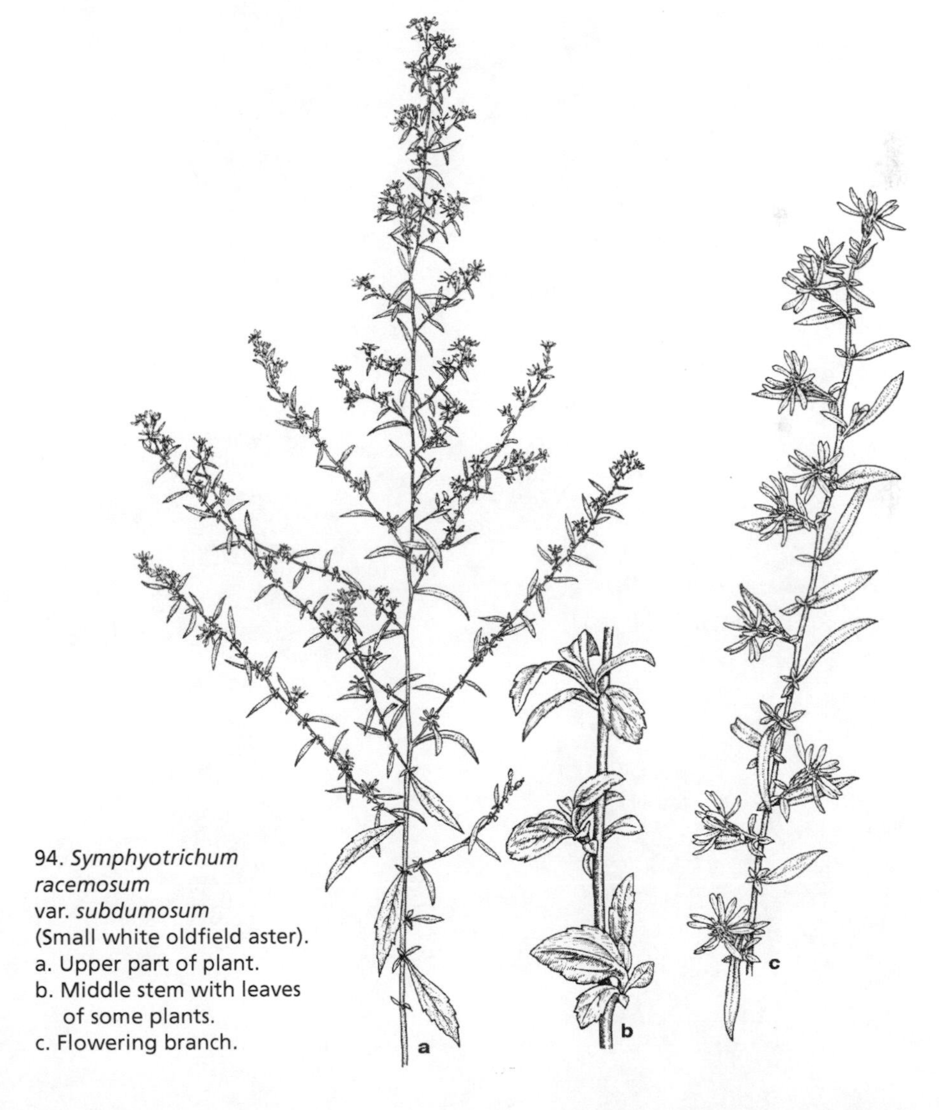

94. *Symphyotrichum racemosum* var. *subdumosum* (Small white oldfield aster).
a. Upper part of plant.
b. Middle stem with leaves of some plants.
c. Flowering branch.

Perennial from long, creeping rhizomes, sometimes colonial, often with a woody caudex; stems 1–3, erect, to 1.0 (–1.5) m tall, glabrous or nearly so, often purple-tinged; basal leaves absent at flowering time, oblanceolate, acute at the apex, tapering to the narrowly winged petiole, crenate-serrate, glabrous or nearly so, to 4 cm long, to 1.5 cm wide; cauline leaves progressively smaller, narrowly lanceolate to linear, acute at the apex, tapering to the sessile and sometimes clasping base, entire or sparsely serrate, glabrous on both surfaces but scabrous above, up to 7 cm long, up to 2 cm wide; inflorescence paniculate with spreading or arching branches, with 50–150 heads on branchlets with bracts linear to acicular, glabrous, to 2 mm long, radiate; involucre 2.5–5.5 mm high, 3–5 mm wide; phyllaries in 4–6 series, unequal, appressed or the outer spreading, linear, linear to narrowly lanceolate, acute at the apex, glabrous, with an elongated green tip more than half as long as

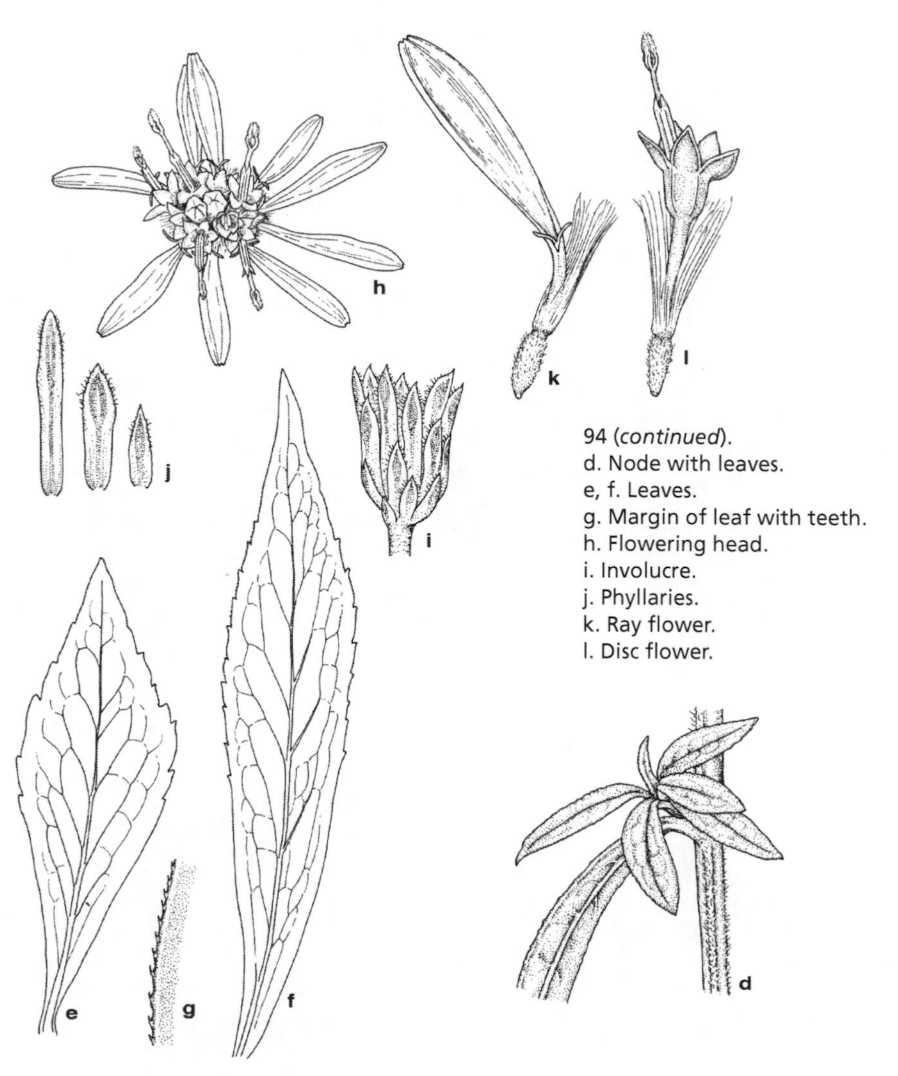

94 (*continued*).
d. Node with leaves.
e, f. Leaves.
g. Margin of leaf with teeth.
h. Flowering head.
i. Involucre.
j. Phyllaries.
k. Ray flower.
l. Disc flower.

the phyllary; receptacle flat, pitted, epaleate; ray flowers 12–20, 3–6 (–8) mm long, pistillate, white, rarely purplish; disc flowers up to 25, 2.5–4.5 mm long, bisexual, cream or yellow, often becoming pinkish; cypselae obovate, more or less flat, sparsely hairy, 4- to 5-nerved, gray to tan, 1.0–1.8 mm long; pappus of numerous white capillary bristles 2.5–3.5 mm long.

Common Name: Small white oldfield aster.
Habitat: Moist open ground.
Range: New Brunswick to Ontario to Iowa, south to Texas and Florida.
Illinois Distribution: Occasional to common throughout the state.

This plant in the past usually has been called *Aster vimineus*, but that name apparently applies to a different species. Our plants belong to var. *subdumosum*, named because of its strong resemblance to *S. dumosum* with its numerous heads on branchlets that bear many tiny bracts. The type specimen for var. *subdumosum* is from Richland County. The green tips of the phyllaries of *S. racemosum* var. *subdumosum* are at least half as long as the phyllary, while the green tips of the phyllaries of *S. dumosum* are less than half the length of the phyllary.

This plant flowers from August to October.

27. **Symphyotrichum lateriflorum** (L.) G. L. Nesom, A. Love & D. Love, Taxon 31:359. 1982. Fig. 95.
*Solidago laterifloru*s L. Sp. Pl. 2:879. 1753.
Aster pendulus Ait. Hort. Kew. 3:204. 1789.
Aster diffusus Ait. Hort. Kew. 3:205. 1789.
Aster miser Nutt. Gen. 2:158. 1818.
Aster horizontalis Desf. Tabl. Ecole Bot. ed. 3, 402. 1829.
Aster hirsuticaulis Lindl. ex DC. Prodr. 5:242. 1836.
Aster miser Nutt. var. *diffusus* (Ait.) Torr. & Gray, Fl. N. Am. 2:130. 1841.
Aster diffusus Ait. var. *bifrons* Gray, Syn. Fl. 1:187. 1884.
Aster lateriflorus (L.) Britt. Trans. N.Y. Acad. Sci. 9:10. 1889.
Aster lateriflorus (L.) Britt. var. *hirsuticaulis* (Lindl.) Porter, Mem. Torrey Club 5:324. 1894.
Aster lateriflorus (L.) Britt. var. *horizontalis* (Desf.) Farw. Asa Gray Bulletin 3:21. 1895.
Aster lateriflorus (L.) Britt. var. *pendulus* (Ait.) E. S. Burgess in Britt. & Brown, Ill. Fl. N.E.U.S. 3:380. 1898.
Symphyotrichum lateriflorum (L.) A. Love & D. Love var. *horizontale* (Desf.) G. L. Nesom, Phytologia 77:285. 1995.

Cespitose perennial from short, stout rhizomes and a thickened caudex; stems 1 to several, erect to less commonly spreading, to 1.5 m tall, glabrous or sparsely pubescent, especially near the tip; basal leaves absent at flowering time, sessile or short-petiolate, oblanceolate to obovate, obtuse to acute at the apex, tapering to the base, entire or serrate, glabrous above, pubescent on the lower surface only on the midvein, to 12 cm long, to 3 cm wide; cauline leaves progressively smaller, sessile, lanceolate to linear, acute at the apex, tapering to the base, entire or shallowly serrate, glabrous except for the pubescent midvein on the lower surface, 5–15 cm long, up to 1.5 cm wide; inflorescence paniculate, the branches ascending to spreading, the bracts to 2 cm long, with 30–100 heads up to 15 mm in diameter, radiate; involucre 4–7 mm high, 4–6 mm wide; phyllaries in 3–5 series, unequal, linear to narrowly oblong, obtuse to acute at the apex, glabrous, with green or purplish tips; receptacle flat, pitted, epaleate; ray flowers 9–15 (–20), up to 7 (–8) mm long, pistillate, white or rarely lavender; disc flowers up to 15 (–20), 2.5–4.5 mm long, the lobes reflexed, bisexual, yellow or purplish; cypselae obovate, flat,

95. *Symphotrichum lateriflorum* (Calico aster).

a. Habit.
b. Upper part of plant.
c. Leaf.
d. Flowering branch.
e. Flowering head.
f. Phyllaries.
g. Ray flower.
h. Disc flower.

sparsely pubescent, 3- to 5-nerved, 1.5–2.2 mm long; pappus of numerous white capillary bristles 2.5–4.5 mm long.

Common Name: Calico aster; one-sided aster.
Habitat: Floodplain woods, calcareous fens, sloughs, along river and streams, around lakes and ponds.
Range: New Brunswick to Manitoba, south to Texas and Florida.
Illinois Distribution: Common throughout the state.

This species is similar in appearance to *S. lanceolatum* since both species have more or less glabrous stems, white flower heads, and leaves tapering to the base. It differs from *S. lanceolatum* by the reflexed corolla lobes of the disc flowers and the pubescent midvein on the lower surface of the leaves. It also resembles *S. ontarionis* by its reflexed corolla lobes of the disc flowers, but the stems of *S. ontarionis* are uniformly pubescent, as are the leaves.

Most specimens in Illinois have erect to spreading stems, although some plants may have widely spreading stems. These may be known as var. *horizontale*, but this variety has not been transferred to *Symphyotrichum*.

The flowers open from August to October.

28. **Symphyotrichum ontarionis** (Wieg.) G. L. Nesom, Phytologia 77:287. 1995. Fig. 96.
Aster diffusus Ait. var. *thyrsoideus* Gray, Syn. Fl. 1:187. 1884.
Aster missouriensis Britt. in Britt. & Brown var. *thyrsoideus* (Gray) Wieg. Rhodora 30:177. 1928.
Aster ontarionis Wieg. Rhodora 30:179. 1928.
Aster panotrichus S. F. Blake, Journ. Wash. Acad. Sci. 21:327. 1931.
Aster panotrichus S. F. Blake var. *thyrsoideus* (Gray) S. F. Blake, Journ. Wash. Acad. Sci. 21: 327. 1931.

Perennial from long, creeping rhizomes, forming colonies; stem usually 1, erect to ascending, up to 80 (–100) cm tall, puberulent throughout with short spreading hairs but becoming glabrate near the base; basal leaves absent at flowering time, sessile or on very short petioles, oblanceolate to obovate, acute at the apex, tapering to the base, serrate, pubescent on both surfaces, to 8 cm long, to 3.5 cm wide; cauline leaves progressively smaller, sessile, lanceolate to oblong to linear, acute at the apex, usually entire, softly pubescent on both surfaces, to 10 (–12) cm long, to 2 cm wide; inflorescence an open panicle with ascending branches, with 20–50 usually secund heads 8–15 mm in diameter, radiate, the bracts leaflike, to 1.2 cm long; involucre 3–6 mm high, 4–8 mm wide; phyllaries in 2–5 series, linear, obtuse or acute at the green tip, glabrous or sparsely pubescent; receptacle flat, pitted, epaleate; ray flowers 10–25, 3–8 mm long, pistillate, usually white, rarely pale blue; disc flowers 20–50, 2.5–4.5 mm long, bisexual, violet-purple; cypselae obovate, flat, puberulent, 3- to 5-nerved, 0.9–2.0 mm long; pappus of numerous white to cream capillary bristles 3.0–4.5 mm long.

Common Name: Ontario aster.
Habitat: Floodplain woods, river terraces.
Range: Quebec to Minnesota to South Dakota, south to Texas, Alabama, and North Carolina.
Illinois Distribution: Common throughout the state.

Symphyotrichum ontarionis resembles *S. lateriflorum* because of its spreading to reflexed lobes of its disc flowers and white ray flowers, but it differs by its colonial habit, its longer rhizomes, and its stems and leaves pubescent throughout.

This species flowers from August to October.

96. *Symphyotrichum ontarionis* (Ontario aster).

a. Upper part of plant.
b. Node with leaf.
c. Flowering head.
d. Phyllary.
e. Ray flower.
f. Disc flower.
g. Cypsela.

29. **Symphyotrichum lanceolatum** (Willd.) G. L. Nesom, Phytologia 77:284. 1995.
Aster lanceolatus Willd. Sp. Pl. 3:2050. 1803.
Aster simplex Willd. Enum. Pl. 2:887. 1809.
Aster tenuifolius L. var. *ramosissimus* Torr. & Gray, Fl. N. Am. 2:132–133. 1841.
Aster simplex Willd. var. *ramosissimus* (Torr. & Gray) Cronq. Bull. Torrey Club 74:145. 1947.
Aster lanceolatus Willd. ssp. *simplex* (Willd.) A. G. Jones, Phytologia 55:383. 1984.

Perennial from long, creeping rhizomes, usually forming colonies; stem solitary, erect, slender or somewhat stout, to 1.5 m tall, glabrous except sometimes with lines of pubescence below each leaf, rarely densely tomentose; basal leaves absent at flowering time, short-petiolate, linear to broadly lanceolate, acute and mucronate at the apex, tapering to the base, serrate-crenate, glabrous, to 80 mm long, to 20 mm wide; cauline leaves linear to lanceolate, acute at the apex, tapering to the sessile or subsessile base, entire or occasionally serrate, glabrous on both surfaces, to 15 cm long, to 3.5 mm wide; inflorescence a panicle of 30–80 often crowded heads, radiate; involucre 3–6 mm high, 3–6 mm wide; phyllaries in 3–6 series, unequal, imbricate, linear, acute, glabrous except for the often ciliate margins, scarious, erose, with a green vein and tip; receptacle flat, pitted, epaleate; ray flowers 20–30 (-50), 4–10 (-12) mm long, pistillate, white, rarely lavender; disc flowers about 40 (-50), bisexual, tubular, purplish; cypselae obovate, flat, strigose, 4- 50 5-nerved, gray or tan, 0.8–1.0 mm long; pappus of numerous white or tawny capillary bristles 5–6 mm long.

Symphyotrichum lanceolatum is similar in appearance to *S. lateriflorum* and *S. ontarionis*, but differs by its erect lobes of the disc flowers, where the other two species have spreading to reflexed lobes of the disc flowers.

Four varieties occur in Illinois:

a. Stems more or less glabrous.
  b. Involucre 3.5–5.5 mm high; heads not crowded.
    c. Leaves linear to narrowly lanceolate, the uppermost much reduced .... 29a. *S. lanceolatum* var. *lanceolatum*
    c. Leaves broadly oblanceolate, the uppermost not much reduced .... 29b. *S. lanceolatum* var. *latifolium*
  b. Involucre 3–4 mm high; heads crowded........... 29c. *S. lanceolatum* var. *interior*
a. Stems densely tomentose .... 29d. *S. lanceolatum* var. *hirsuticaule*

29a. **Symphyotrichum lanceolatum** (Willd.) G. L. Nesom var. **lanceolatum**. Fig. 97.
Aster lanceolatus Willd. Sp. Pl. 3:2050. 1803.
Aster simplex Willd. Enum. Pl. 2:887. 1809.
Aster tenuifolius L. var. *ramosissimum* Torr. & Gray, Fl. N. Am. 2:132–133. 1841.
Aster paniculatus Lam. var. *acutidens* E. S. Burgess in Britt. & Brown, Ill. Fl. N.E.U.S. 3:378. 1898.
Aster acutidens (E. S. Burgess) Smyth, Trans. Kansas Acad. Sci. 16:160. 1899.
Aster simplex Willd. var. *ramosissimum* (Torr. & Gray) Cronq. Bull. Torrey Club 74:145. 1947.

Stems more or less glabrous; involucre 3.5–5.5 mm high; heads not crowded; leaves linear to narrowly lanceolate, the uppermost much reduced.

Common Name: White panicled aster.
Habitat: Wet meadows, wet ditches, moist ground.
Range: Labrador to Saskatchewan, south to Texas, Louisiana, Kentucky, and Virginia.
Illinois Distribution: Common throughout the state.

This is one of the more common varieties of *S. lanceolatum* in Illinois.

This variety flowers from August to October.

97. *Symphyotrichum lanceolatum* var. *lanceolatum* (White panicled aster).
a. Upper part of plant.
b. Flowering head.
c. Phyllaries.
d. Ray flower.
e. Disc flower.
f. Cypsela.

29b. **Symphyotrichum lanceolatum** (Willd.) G. L. Nesom var. **latifolium** (Semple & Chmielewski) G. L. Nesom, Phytologia 77:285. 1995. Fig. 97.

Aster lanceolatus Willd. var. *latifolius* Semple & Chmielewski, Can. Journ. Bot. 65:1060. 1987.

Stems more or less glabrous; involucre 3.5–5.5 mm high; heads not crowded; leaves oblanceolate, the uppermost not very much reduced.

Common Name: Broad-leaved white panicled aster.
Habitat: Moist soil.
Range: Maine to Manitoba, south to Texas and Florida.
Illinois Distribution: Scattered throughout Illinois.

This variety occurs occasionally in all parts of the state. It flowers from August to October.

97 (*continued*).
Symphyotrichum lanceolatum var. *latifolium* (Broad-leaved white panicled aster).
g. Upper part of plant.
h. Flowering head.
i. Involucre.
j. Ray flower.
k, l. Disc flowers.
m. Cypsela.

29c. **Symphyotrichum lanceolatum** (Willd.) G. L. Nesom var. **interior** (Wieg.) G. L. Nesom, Phytologia 77:284. 1995. Fig. 97.

Aster interior Wieg. Rhodora 35:35. 1933.

Aster simplex Willd. var. *interior* (Wieg.) Cronq. Bull. Torrey Club 74:146. 1947.

Aster lanceolatus Willd. ssp. *interior* (Wieg.) A. G. Jones, Phytologia 55:383. 1984.

Stems more or less glabrous; involucre 3–4 mm high; heads crowded.

Common Name: Small-flowered white panicled aster.

Habitat: Bottomland forests.

Range: Quebec and Ontario to Nebraska, south to Oklahoma, Arkansas, Kentucky, and Pennsylvania.

Illinois Distribution: Scattered in the southern half of Illinois.

This variety looks quite different from var. *lanceolatum* and var. *latifolium* because of its crowded smaller flowering heads. It appears to be restricted to the southern half of the state, where it occurs primarily in bottomland forests. It flowers from August to October.

97 (*continued*). *Symphyotrichum lanceolatum* var. *interior* (Small-flowered white panicled aster). n, o. Habit. p. Node with leaf. q. Flowering head. r. Phyllary. s. Ray flower. t. Disc flower.

29d. **Symphyotrichum lanceolatum** (Willd.) G. L. Nesom var. **hirsuticaule** (Semple & Chmielewski) G. L. Nesom, Phytologia 77:284. 1995.

Aster lanceolatus Willd. var. *hirsuticaulis* Semple & Chmielewski, Can. Journ. Bot. 65:1058. 1987.

Stems densely tomentose.

Common Name: Hairy white panicled aster.

Habitat: Moist soil, mostly on glacial deposits.

Range: Ontario to Manitoba, south to Iowa, Illinois, and Michigan.

Illinois Distribution: Restricted to a few stations in the northernmost counties of the state.

Although the flowering heads are definitely those of *S. lanceolatum*, the tomentose stems give this variety a very distinct appearance, more like that of *S. ontarionis*.

The occurrence of this variety primarily on glacial deposits is unique for this variety. It flowers from August to October.

30. **Symphyotrichum praealtum** (Poir.) G. L. Nesom, Phytologia 77:289. 1995.

Aster praealtus Poir. Encycl., Suppl. 1:493. 1811.

Perennial from long, creeping rhizomes, often colonial; stem usually 1, erect to ascending, to 1.5 m tall, densely pubescent, at least on the upper half, but usually glabrate near the base; basal leaves absent at flowering time, oblanceolate, obtuse to acute at the apex, tapering to the sessile or short-petiolate base, entire, glabrous or scabrous above, glabrous or nearly so below, conspicuously reticulate-veined, to 7 cm long, to 2.5 cm wide; cauline leaves progressively smaller, linear-lanceolate to narrowly elliptic, acute at the apex, rounded or tapering to the sessile base, entire or sparingly dentate, glabrous but sometimes scabrous above, glabrous or puberulent beneath, conspicuously reticulate-veined beneath, to 15 cm long, to 2 cm wide; inflorescence paniculate, with loosely ascending to spreading branches, with 1 to a few radiate heads 1.5–2.5 cm in diameter, with narrow bracts to 15 mm long; involucre 4–8 mm high, 6–9 mm wide; phyllaries in 4 or 5 series, unequal, linear to narrowly oblong, obtuse to acute at the apex, glabrous or sparsely pubescent, greenish or reddish at the tip; receptacle flat, pitted, epaleate; ray flowers 20–35, 5–15 mm long, pistillate, blue; disc flowers up to 35, 4–6 mm long, bisexual, tubular, yellow; cypselae obovate, flat, pubescent, 4- to 5-nerved, 1–2 mm long, purplish; pappus of numerous white capillary bristles 4.0–6.5 mm long.

This species is distinguished by its conspicuously reticulate-veined leaves. It is somewhat variable, and three varieties may be recognized in Illinois.

a. Leaves rather thick, firm, lanceolate to elliptic.
  b. Leaves more or less smooth above . . . . 30a. *S. praealtum* var. *praealtum*
  b. Leaves harshly scabrous above. .30b. *S. praealtum* var. *subasper*
a. Leaves thin, linear to linear-lanceolate . . . . .30c. *S. praealtum* var. *angustior*

30a. **Symphyotrichum praealtum** (Poir.) G. L. Nesom var. **praealtum**
Aster praealtum Poir. Encycl., Suppl. 1:493. 1811.

Leaves rather thick, firm, lanceolate to elliptic, more or less smooth above.

Common Name: Net-veined aster.
Habitat: Wet meadows, calcareous fens, moist soil.
Range: Maine to Ontario and South Dakota, south to Texas and Florida.
Illinois Distribution: Occasional to common in the northern three-fourths of Illinois.

This is the more common variety in Illinois. It flowers from August to October.

30b. **Symphyotrichum praealtum** (Poir.) G. L. Nesom var. **subasper** (Lindl.) G. L. Nesom, Phytologia 77:289. 1995.
Aster subasper Lindl. Comp. Bot. Mag. 1:97. 1835.
Aster salicifolius Ait. var. *subasper* (Lindl.) Gray, Syn. Fl. 1:188. 1884.
Aster praealtus Poir. var. *subasper* (Lindl.) Wieg. Rhodora 35:25–26. 1933.

Leaves rather thick, firm, lanceolate to elliptic, harshly scabrous above.

Common Name: Veiny aster.
Habitat: Moist soil.
Range: Maine to Ontario and South Dakota, south to Texas and Florida.
Illinois Distribution: Occasional in the northern half of the state.

The scabrous upper leaf surface distinguishes this variety. It flowers from August to October.

30c. **Symphyotrichum praealtum** (Poir.) G. L. Nesom var. **angustior** (Wieg.) G. L. Nesom, Phytologia 77:289. 1995. Fig. 98.
Aster praealtus Poir. var. *angustior* Wieg. Rhodora 35:24. 1933.

Leaves thin, linear to linear-lanceolate.

Common Name: Willow-leaved aster.
Habitat: Wet meadows, moist soil.
Range: Maine to Ontario and South Dakota, south to Texas and Florida.
Illinois Distribution: Occasional in the northern three-fourths of Illinois.

The thin, narrow leaves give this variety a distinct appearance. It flowers from August to October.

31. **Symphyotrichum boreale** (Torr. & Gray) A. Love & D. Love, Taxon 31:358. 1982. Fig. 99.
Aster latifolius Lindl. var. *borealis* Torr. & Gray, Fl. N. Am. 2:138. 1841.
Aster borealis (Torr. & Gray) Prov. Fl. Canad. 1:308. 1862.
Aster junciformis Rydb. Bull. Torrey Club 37:142. 1910.

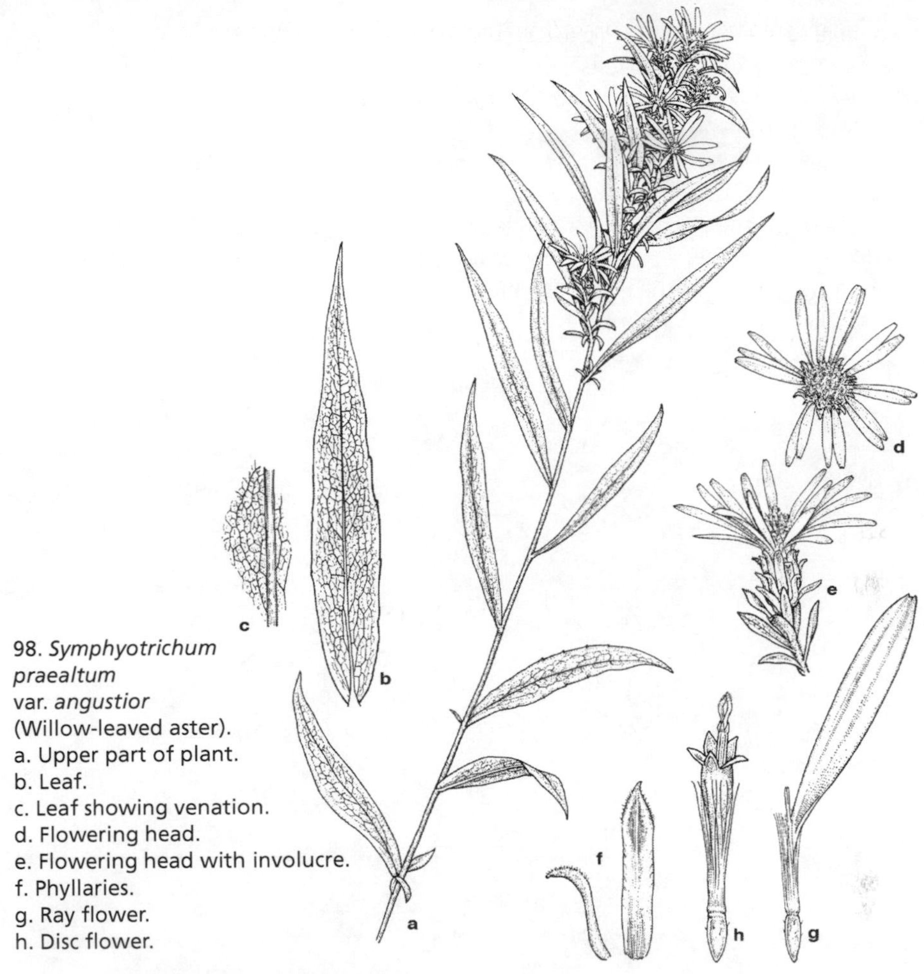

98. *Symphyotrichum praealtum* var. *angustior* (Willow-leaved aster).
a. Upper part of plant.
b. Leaf.
c. Leaf showing venation.
d. Flowering head.
e. Flowering head with involucre.
f. Phyllaries.
g. Ray flower.
h. Disc flower.

Perennial from long, thin rhizomes; stems 1 to a few, erect, slender, to 80 cm tall, glabrous in the lower half, sometimes pubescent in the upper half, distinctly grooved, often reddish, at least at the base; basal leaves absent at flowering time, broadly elliptic, acute and mucronulate at the tip, tapering to the winged petiolate base, usually entire, glabrous; cauline leaves narrowly lanceolate to elliptic, acute to acuminate at the tip, mucronulate, tapering to the sessile or often clasping base, entire but often scabrous along the margins, glabrous except for the pubescent midvein, to 15 cm long, to 5.5 (–6.0) mm wide; inflorescence an open corymb, the branches ascending, with 5–20 radiate heads, sometimes with a solitary head; involucre hemispheric, 5–8 mm high, up to 10 mm wide; phyllaries in 3–5 series, appressed, linear to lanceolate, acute, glabrous, often with a purplish tip and purplish, often ciliate margin; receptacle flat, pitted, epaleate; ray flowers (20–) 30–40, 7–15 mm long, pistillate, white to lavender; disc flowers about 30 (–40), bisexual, tubular, yellow-brown to pale yellow, 3.5–6.5 mm long; cypselae flat, obovate, glabrous or strigose, yellow to purple-brown, several-nerved, 1–2 mm long; pappus of numerous white capillary bristles 2.5–6.5 mm long.

Common Name: Northern bog aster; rush aster.
Habitat: Bogs, calcareous fens.
Range: Nova Scotia to Alaska, south to Washington, northern Colorado, northern Illinois, and West Virginia.
Illinois Distribution: Occasional in northeastern Illinois, south to Peoria County.

This species is distinguished by its relatively short heads, narrow leaves up to 6 mm wide, few flower heads per inflorescence, and usually more than 30, often white, ray flowers. Some of the leaves are often clasping.

Rydberg's *Aster junciformis* is a synonym for this species.

Symphyotrichum boreale flowers from August to October.

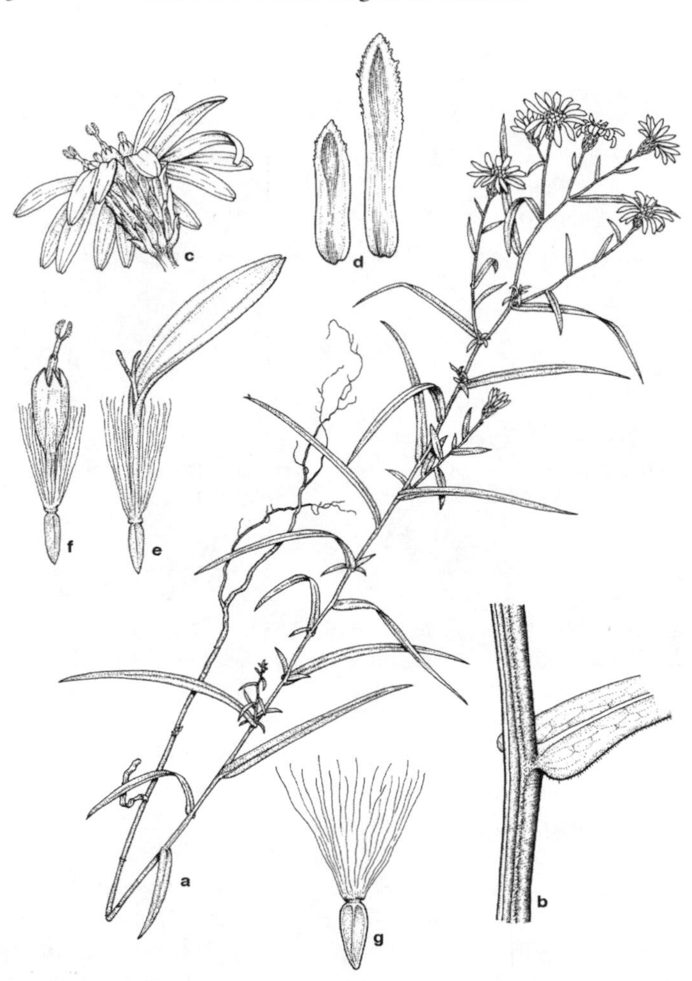

99. *Symphyotrichum boreale* (Northern bog aster).

a. Habit.
b. Node with base of leaf.
c. Flowering head.
d. Phyllaries.
e. Ray flower.
f. Disc flower.
g. Cypsela.

32. **Symphyotrichum puniceum** (L.) A. Love & D. Love, Taxon 31:359. 1982. Fig. 100.

Aster puniceus L. Sp. Pl. 2:875. 1753.

Perennial from short, stout or long creeping rhizomes and a thickened caudex; stems 1 to several, stout, more or less erect, to 2.5 m tall, hispid, often becoming glabrate near the base, usually purplish, at least at the nodes; basal leaves absent at flowering time, oblanceolate to lanceolate, acute to acuminate and mucronate at the apex, tapering to the broad-leaved sessile base or auriculate-clasping, remotely toothed or entire, hispid, to 10 cm long, to 2 cm wide; cauline leaves narrowly lanceolate to ovate, acute at the apex, auriculate-clasping at the base, entire to

100. *Symphyotrichum puniceum* (Purplestem aster).

a. Upper part of plant.
b. Base of plant with roots.
c. Flowering head.
d. Phyllaries.
e. Ray flower.
f. Disc flower.
g. Cypsela.

irregularly serrate, scabrous on the upper surface, glabrous or pilose on the lower surface, to 15 cm long, to 4 cm wide; inflorescence paniculate, with spreading to ascending branches with 30–50 radiate heads 1.5–2.5 cm in diameter, the leaflike bracts up to 15 mm long; involucre campanulate, 5–12 (–15) mm high, 7–14 mm wide; phyllaries in 3–6 series, more or less equal, linear to narrowly oblanceolate, acute to acuminate, glabrous or nearly so except for the ciliate margins, hyaline, erose; receptacle flat, pitted, epaleate; ray flowers 20–60, 7–20 mm long, pistillate, dark blue to purple; disc flowers 30–50 (–50), 4.0–6.5 mm long, bisexual, tubular, yellow but sometimes becoming pink or purple; cypselae oblanceolate, flat, pubescent, becoming glabrate, 3- to 4-nerved, purple to brown, 2–4 mm long; pappus of numerous white capillary bristles 4–6 mm long.

Common Name: Purplestem aster; swamp aster.
Habitat: Calcareous fens, moist soil.
Range: Labrador to British Columbia, south to Nebraska, Missouri, Alabama, and Georgia.
Illinois Distribution: Occasional in the northern three-fourths of the state.

This blue-flowered species is distinguished by its auriculate-clasping leaves that are scabrous. The stems are hispid.

Symphyotrichum puniceum flowers from August to October.

33. **Symphyotrichum firmum** (Nees) G. L. Nesom, Phytologia 77:282. 1995. Fig. 101.
Aster firmus Nees, Syn. Aster. Herb. 25. 1818.
Aster puniceus L. var. *firmus* (Nees) Torr. & Gray, Fl. N. Am. 2:131. 1841.
Aster puniceus L. var. *lucidulus* Gray, Syn. Fl. N. Am. 1:195. 1884.
Aster lucidulus (Gray) Wieg. Rhodora 26:4. 1924.

Perennial from long, creeping rhizomes, often forming colonies; stem usually solitary, erect, to 2.5 m tall, glabrous except sometimes for a line of pubescence below the leaves; leaves firm, absent at flowering time, oblanceolate, acute at the apex, tapering to very short, winged petioles, entire to crenulate, revolute, glabrous or nearly so, to 10 cm long, to 2 cm wide; cauline leaves lanceolate to ovate-lanceolate, acute at the tip, sessile and clasping at the base, entire or nearly so, glabrous on the upper surface, glabrous or pubescent on the midnerve beneath, 4–7 cm long, 1.0–2.5 (–3.5) cm wide; inflorescence an open panicle, the branches ascending, with 60–100 radiate heads and 4–6 linear bracts; involucre 6–12 mm high, 6–10 mm wide; phyllaries in 4 or 5 series, imbricate, linear, long-attenuate, loosely arranged, scarious, erose, glabrous except for the ciliate margins, with elongated green tips; receptacle flat, pitted, epaleate; ray flowers up to 40, 10–18 mm long, 1.0–1.2 mm wide, pistillate, blue, rarely white; disc flowers up to 50, bisexual, tubular, yellow, becoming pink or purple, 5.0–6.5 mm long; cypselae obovate, flat, glabrous or nearly so, purple or brown, 3- to 4-nerved, 1.5–3.0 mm long; pappus of numerous white capillary bristles 5–8 mm long.

101. *Symphyotrichum firmum* (Glossy-leaved aster).

a. Upper part of plant.
b. Rhizome.
c. Leaf and node.
d. Flowering head.
e. Phyllaries.
f. Ray flower.
g. Disc flower.
h. Cypsela.

Common Name: Glossy-leaved aster; swamp aster.
Habitat: Wet prairies, calcareous fens.
Range: Ontario to Minnesota and Saskatchewan, south to Nebraska, Missouri, and Illinois; also in Georgia.
Illinois Distribution: Occasional in the northern half of the state.

The distinguishing features of this species are its elongated rhizomes, clasping cauline leaves, eglandular bracts, and stems that are glabrous or with a fine line of pubescence. The somewhat similar *S. puniceum* has short rhizomes and hispid stems. The flowers are occasionally white.

Symphyotrichum firmum flowers from August to October.

34. **Symphyotrichum prenanthoides** (Muhl. ex Willd.) G. L. Nesom, Phytologia 77: 290. 1995. Fig. 102.
Aster prenanthoides Muhl. ex Willd. Sp. Pl. 3:2046. 1803.

Perennial from long, creeping rhizomes, colonial; stems 1 to a few, erect to ascending, often zigzag, to 1.2 m tall, glabrous or with lines of pubescence beneath each leaf; basal leaves absent at flowering time, ovate to lanceolate, acuminate and mucronate at the apex, tapering or abruptly contracted to a broadly winged petiole that frequently clasps the stem, glabrous and scabrous on the upper surface, glabrous or villous on the lower surface, entire or usually coarsely serrate, up to 8 cm long, up to 2.5 cm wide; cauline leaves progressively smaller, elliptic-lanceolate to ovate, acuminate at the apex, tapering to the sessile or broadly winged petioles, sharply serrate, glabrous or villous, to 20 cm long, to 5 cm wide; inflorescence a broad, more or less flat-topped panicle with 10–50 radiate heads, the bracts lanceolate, to 12 mm long; involucre turbinate, 5–7 mm high, about as wide; phyllaries in 4–6 series, spreading to more or less squarrose, linear-lanceolate to oblong-lanceolate, obtuse to acute, subulate, narrowly hyaline, scarious, erose, glabrous or with ciliate margins; receptacle flat, pitted, epaleate; ray flowers 15–30, 7–15 mm long, pistillate, blue or purple, rarely white; disc flowers up to 50 (–65), 3.5–5.0 mm long, bisexual, tubular, yellow but usually turning brown or purple; cypselae oblanceolate to obovate, more or less flat, strigose, 4- to 6-nerved, 2.0–3.5 mm long; pappus of yellowish capillary bristles 3.5–4.5 mm long.

Common Name: Crookedstem aster.
Habitat: Moist ground, north-facing slopes, swamps.
Range: Connecticut to Ontario to Minnesota, south to Iowa, Illinois, Kentucky, and North Carolina.
Illinois Distribution: Rare in the northern two-fifths of Illinois; also Jackson County.

This is the only species of *Symphyotrichum* in Illinois with a more or less flat-topped inflorescence. The sharply serrate, acuminate, clasping leaves are also distinctive.

This species flowers from August to October.

102. *Symphyotrichum prenanthoides* (Crookedstem aster).

a. Upper part of plant.
b. Leaf.
c. Flowering head.
d. Involucre.
e. Phyllary.
f. Ray flower.
g. Disc flower.
h. Cypsela.

103. *Symphyotrichum turbinellum* (Top-shaped aster).

a. Upper part of plant.
b. Leaf.
c. Flowering head with involucre.
d. Phyllaries.
e. Ray flower.
f. Disc flower.
g. Cypsela.

35. **Symphyotrichum turbinellum** (Lindl.) G. L. Nesom, Phytologia 77:293. 1995. Fig. 103.

Aster turbinellus Lindl. Comp. Bot. Mag. 1:98. 1835.

Perennial from a thickened caudex; stems 1 to several, erect to ascending to spreading, to 1.0 (–1.5) m tall, glabrous or hirsute; basal leaves absent at flowering time, oblanceolate to narrowly oblong, obtuse to acute at the apex, rounded at the sessile or short-petiolate base, entire to sparsely toothed, scabrous or glabrous, to 12 cm long, to 2 cm wide; cauline leaves progressively smaller, linear to linear-lanceolate to narrowly elliptic, acute at the apex, tapering to the sessile base, entire, glabrous but scabrous and sometimes ciliate on the margins, up to 10 cm long, up to 1.8 cm wide; inflorescence a panicle of 20–50 radiate heads, the heads 2–3 cm across, the branches with linear to lanceolate bracts up to 12 mm long; involucre 7–12 mm high, turbinate, about as wide; phyllaries in 6–9 series, unequal, linear to oblanceolate, obtuse, with white borders, loose and often squarrose, glabrous; receptacle flat, pitted, epaleate; ray flowers 15–20, 10–18 mm long, pistillate, blue to purple; disc flowers up to 30, 4.5–6.5 mm long, bisexual, tubular, yellow; cypselae obovate, more or less flat, strigose, 2- to 4-nerved, 2–3 mm long, light brown to gray; pappus of numerous white to reddish brown capillary bristles 4.5–6.5 mm long.

Common Name: Top-shaped aster; prairie aster.
Habitat: Dry woods, prairies.
Range: Illinois to Nebraska, south to Oklahoma and Louisiana.
Illinois Distribution: Occasional in the southern three-fifths of the state.

This species is distinguished by its large top-shaped involucres with phyllaries in 6–9 series. It flowers during September and October.

20. **Brachyactis** Ledeb.—Rayless Aster

Annual herbs; stems 1 to a few, erect to ascending, glabrous or nearly so; basal leaves absent at flowering time; cauline leaves alternate, all similar, linear to linear-lanceolate, entire, sessile, glabrous or nearly so; inflorescence a small panicle or raceme, with few heads that are rayless or nearly so; heads seemingly with only disc flowers, although minute ray flowers are actually present; involucre campanulate; phyllaries in 2–5 series, more or less equal, linear to linear-oblong, green; receptacle flat, pitted, epaleate; ray flowers present but minute in several series, pinkish, pistillate; disc flowers yellow, tubular, bisexual; cypselae narrowly ellipsoid, not flat, 2- or 3-nerved, pubescent; pappus of numerous white capillary bristles in 1 series.

I am recognizing *Brachyactis* separate from *Symphyotrichum* because of the minute ray flowers in numerous series, its annual habit, and its cauline leaves that are all similar.

This genus consists of five species, mostly in the western United States and northern Asia.

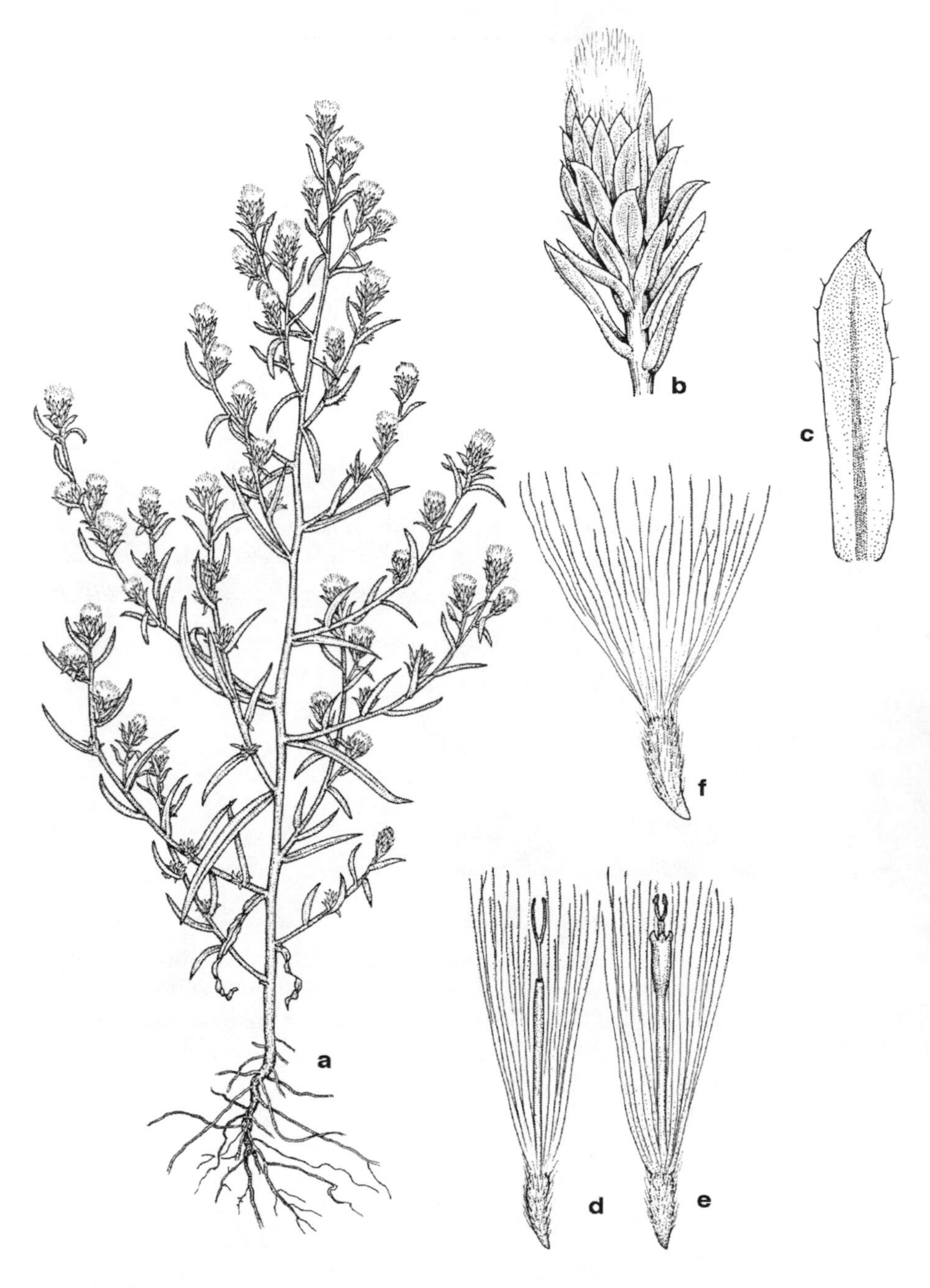

104. *Brachyactis ciliata* (Rayless aster).

a. Habit.
b. Flowering head.
c. Phyllary.
d, e. Disc flowers.
f. Cypsela.

1. **Brachyactis ciliata** (Ledeb.) Ledeb. Fl. Ross. 2:495. 1846. Fig. 104.
Erigeron ciliatus Ledeb. Icon. Pl. 1:24. 1829.
Tripolium angustum Lindl. f. Bor. Am. 2:15. 1834.
Aster angustus (Lindl.) Torr. & Gray, Fl. N. Am. 2:162. 1842, *non* Nees (1818).
Brachyactis angusta (Lindl.) Britt. Ill. Fl. N. U.S. 3:383. 1898.
Aster brachyactis S. F. Blake, Contr. U. S. Natl. Herb. 25:564. 1925.
Brachyactis ciliata (Ledeb.) Ledeb. ssp. *angusta* (Lindl.) A. G. Jones, Phytologia 55:376. 1984.

Annuals from taproots; stems ascending to erect, usually branched, more or less succulent, glabrous except for a little pubescence in the upper half of the stem, often blue-green, to 75 cm tall; basal leaves spatulate, petiolate, withered by flowering time; cauline leaves alternate, somewhat succulent, linear, acute at the apex, tapering to the sessile base, entire to serrulate, glabrous but with marginal cilia, to 8 (–10) cm long, to 4 (–8) mm wide; heads rayless or nearly so, several in a dense panicle or raceme on glabrous peduncles up to 1 cm long, subtended by narrow foliaceous bracts; involucre narrowly campanulate, 5–10 mm high; phyllaries in 2–4 series, more or less equal, linear to linear-oblong, acute at the apex, green with a narrow scarious margin, glabrous except for marginal cilia; receptacle flat, pitted, epaleate; ray flowers absent or reduced to white or pinkish rays up to 2 mm long, pistillate; disc flowers numerous, tubular, bisexual, white or pinkish, with 4 or 5 shallow lobes; cypselae more or less compressed, gray or purple or purple-streaked, 2- to 4-nerved, strigose, 1.5–2.5 mm long; pappus of numerous white or pinkish capillary bristles 4–6 mm long.

Common Name: Rayless aster.
Habitat: Disturbed soil, particularly along highways that have been heavily salted during the winter.
Range: Throughout the upper two-thirds of the United States, but native only from the Great Plains to the Pacific Ocean.
Illinois Distribution: Confined to the northeastern counties of the state.

This species, native west of Illinois, is sometimes placed in *Symphyotrichum*, but it differs by its nearly rayless flowering heads and its annual habit.

The type specimen of *Brachyactis ciliata,* which has leaves broader than those of our plants, has sometimes been considered a separate species (*B. angusta*) or a variety (var. *angusta*).

This species flowers from July to October.

Tribe Anthemideae

Annuals, biennials, perennials, or shrubs, often aromatic; leaves basal or cauline or both, the cauline alternate, entire, dentate, pinnately lobed, or pinnately compound; heads radiate or discoid, borne singly or in panicles, racemes, corymbs, or spikes; involucre hemispheric or campanulate; phyllaries in 3 or more series, equal or unequal, usually with a scarious margin; receptacle flat or conic, paleate or epaleate; ray flowers in 1 or 2 series, pistillate, usually yellow or white, sometimes absent; disc flowers bisexual, tubular, usually yellow or gray; cypselae obovoid or columnar or prismatic, sometimes wing-angled, sometimes flattened; pappus absent or a crown of minute scales.

Key to the Genera of Anthemideae in Illinois

1. Ray flowers absent.
  2. Pappus absent; cypselae faintly ribbed; receptacle paleate; disc flowers usually gray . . . . 24. *Artemisia*
  2. Pappus a crown of minute scales; cypselae usually strongly ribbed; receptacle epaleate; disc flowers usually yellow.
    3. Cypselae angular or with 5 or 10 strong ribs; plants not aromatic . . . . 21. *Tanacetum*
    3. Cypselae with 3 strong ribs or 5 faint ribs; plants aromatic . . . . 26. *Matricaria*
1. Ray flowers present.
  4. Ray flowers yellow . . . . 27. *Cota*
  4. Ray flowers white.
    5. Rays up to 8 in number, 3–4 mm long; cypselae flat, winged; disc flowers gray . . . . 22. *Achillea*
    5. Rays more than 8 in number, more than 4 mm long; cypselae terete or angular or prismatic, unwinged; disc flowers yellow or greenish yellow.
      6. Plants aromatic.
        7. Cypselae 4-angled; receptacle paleate . . . . 25. *Anthemis*
        7. Cypselae terete or subterete; receptacle epaleate.
          8. Cypselae with 5–10 conspicuous ribs; pappus a crown of minute scales . . . . 21. *Tanecetum*
          8. Cypselae with 5 faint ribs; pappus absent or a crown of minute scales.
            9. Receptacle paleate . . . . 25. *Anthemis*
            9. Receptacle epaleate . . . . 26. *Matricaria*
      6. Plants not aromatic.
        10. Rays up to 20 in number, 7–10 mm long; receptacle paleate . . . . 23. *Chamaemelum*
        10. Rays up to 35 in number, 12–30 mm long; receptacle epaleate.
          11. Cypselae strongly 10-ribbed; pappus absent . . . . 29. *Leucanthemum*
          11. Cypselae strongly 3-ribbed; pappus absent or a crown or minute scales . . . . 28. *Tripleurospermum*

21. **Tanacetum** L.—Tansy

Mostly perennial herbs from rhizomes; leaves basal or basal and cauline, usually pinnately divided; heads usually radiate, usually several in corymbs; involucre hemispheric; phyllaries in 3–5 series, unequal, the margins scarious; receptacle epaleate; ray flowers up to 20, or absent, pistillate, fertile or sterile, usually yellow; disc flowers numerous, tubular, bisexual, yellow, fertile, 4- or 5-lobed; cypselae several-ribbed, gland-dotted, usually glabrous; pappus usually a short crown of scales.

There are more than 150 species in this genus, which is now considered to contain some species previously assigned to *Chrysanthemum* and *Balsamita*.

Three species occur in Illinois.

1. Leaves 1- to 3-pinnately lobed, with at least 3 pairs of lobes, the lobes entire or dentate.
   2. Leaves 1- to 2-pinnately lobed, with 3–5 pairs of lobes, puberulent beneath; pappus none or with a crown of scales 0.1–0.2 mm long . . . . 1. *T. parthenium*
   2. Leaves 2- to 3-pinnately lobed, with 4 or more pairs of lobes, glabrous or sparsely pubescent beneath; pappus a crown of scales up to 0.5 mm long . . . . . . .2. *T. vulgare*
1. Leaves not pinnately lobed, or with a small pair of lobes near the base, the lobes more or less crenate . . . . 3. *T. balsamita*

1. **Tanacetum parthenium** L. Sp. Pl. 2:845. 1753. Fig. 105.
Matricaria parthenium L. Sp. Pl. 2:890. 1753.
Chrysanthemum parthenium (L.) Bernh. Syst. Verz. 145. 1800.
Leucanthemum parthenium (L.) Gren. & Godr. Fl. Fr. 2:145. 1850.

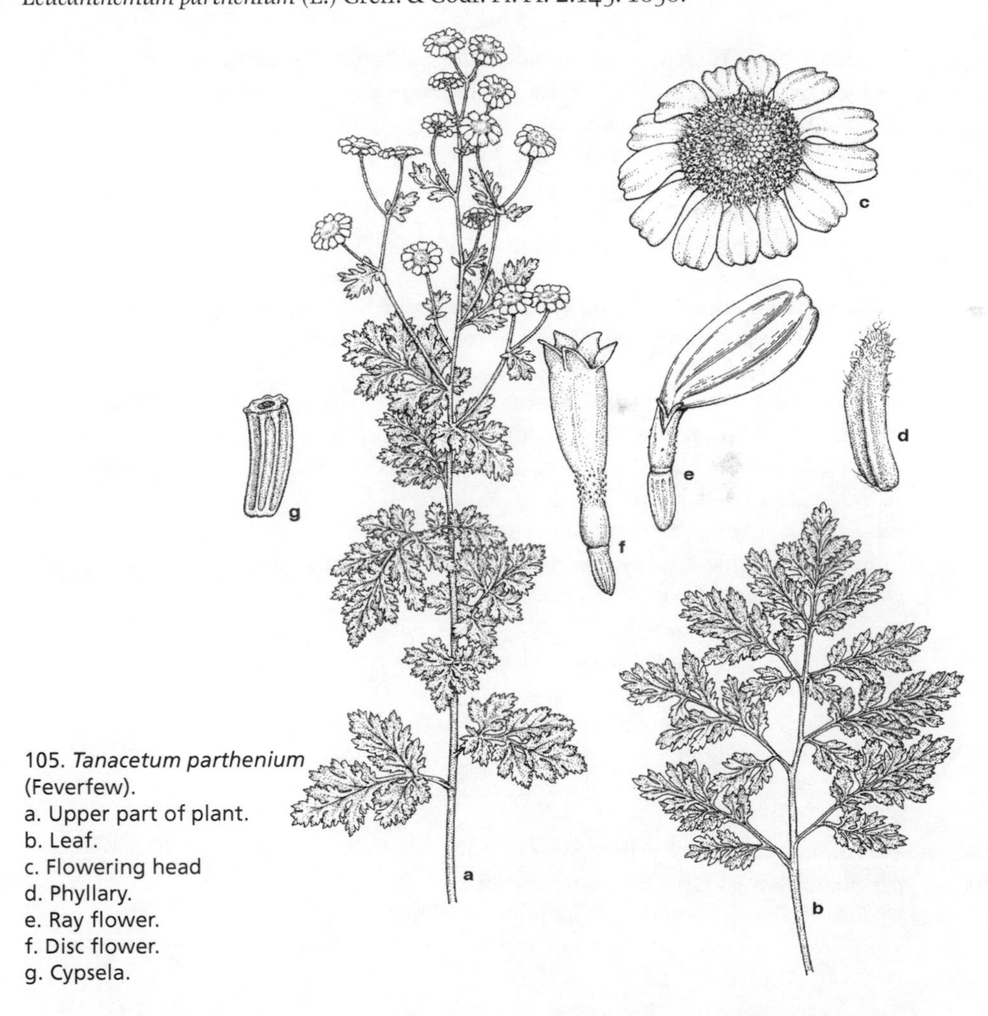

105. *Tanacetum parthenium* (Feverfew).
a. Upper part of plant.
b. Leaf.
c. Flowering head
d. Phyllary.
e. Ray flower.
f. Disc flower.
g. Cypsela.

Perennial herb from rhizomes; stems erect, branched, glabrous or puberulent, to 75 cm tall; leaves alternate, 1- to 2-pinnately divided, with 3 or 5 lobes, the lobes entire or dentate, puberulent beneath, gland-dotted, the lowest petiolate, the upper sessile; heads up to 20 (–30), to 2 cm across, radiate, in corymbs; involucre hemispheric; phyllaries in 3–5 series, unequal, lanceolate, acute at the apex, keeled, pubescent, with a scarious margin; receptacle epaleate; ray flowers up to 20, white, notched at the tip, to 10 mm long, pistillate; disc flowers numerous, tubular, bisexual, yellow, 1.5–2.0 mm high; cypselae with 5–10 ribs, 1–2 mm long, glabrous; pappus a crown of scales 0.1–0.2 mm long, or absent.

Common Name: Feverfew.
Habitat: Disturbed areas.
Range: Native to Europe and Africa; sparingly escaped from cultivation in the United States.
Illinois Distribution: Occasional in the northern half of the state.

For many years, this species was called *Chrysanthemum parthenium*.

Its distinguishing characteristics are its yellow rays, its 1- or 2-pinnately lobed leaves, and its very short or sometimes absent pappus.

Tanacetum parthenium flowers from June to September.

2. **Tanacetum vulgare** L. Sp. Pl. 2:844. 1753. Fig. 106.
Tanacetum vulgare L. var. *crispum* L. Sp. Pl. 2:845. 1753.

Perennial herb; stems erect, mostly unbranched, glabrous or sparsely pubescent, to 1 m tall; leaves of two kinds: basal leaves withered at anthesis; cauline leaves alternate; all leaves 2- or 3-pinnately lobed, with 4 or more pairs of lobes, the lobes dentate, glabrous or sparsely pubescent beneath; heads numerous, crowded in corymbs, to 1.2 cm across, discoid; involucre hemispheric; phyllaries in 3–5 series, unequal, oblong to lanceolate, obtuse or acute at the apex, puberulent or ciliate, with a scarious margin; receptacle epaleate; ray flowers absent; disc flowers numerous, tubular, bisexual, yellow, the outer ones 3- or 4-notched at the tip, the central ones 2–3 mm long; cypselae 4- or 5-angled, 1–2 mm long, gland-dotted; pappus a crown of scales up to 0.5 mm long.

Common Name: Common tansy; golden buttons.
Habitat: Disturbed soil.
Range: Native to Europe and Asia; escaped from cultivation in much of the United States and Canada.
Illinois Distribution: Occasional throughout the state.

This handsome species was popular in gardens at one time. Its disclike heads have peripheral flowers different from the central ones.

Tanacetum vulgare flowers from July to October.

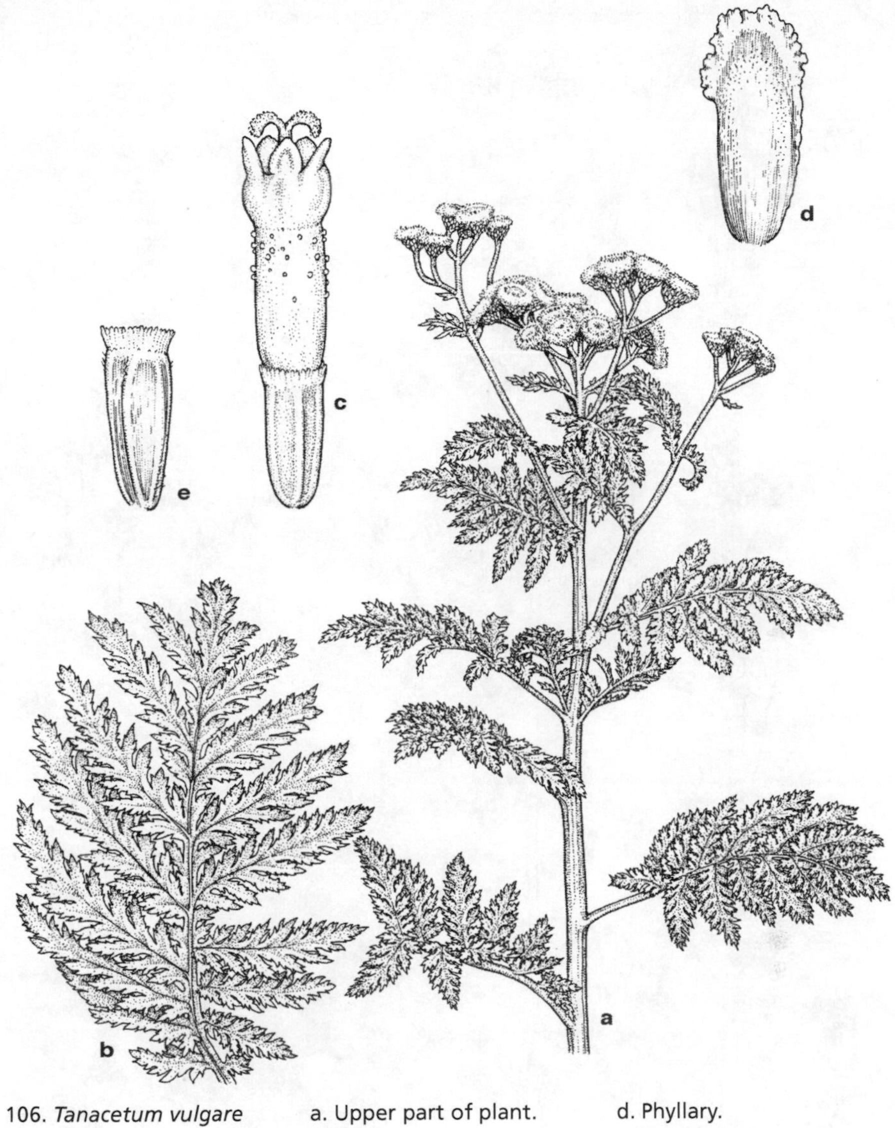

106. *Tanacetum vulgare* (Common tansy).
a. Upper part of plant.
b. Leaf.
c. Disc flower.
d. Phyllary.
e. Cypsela.

3. **Tanacetum balsamita** L. Sp. Pl. 2:845. 1753. Fig. 107.
Chrysanthemum balsamita (L.) Sp. Pl., ed. 2, 2:1252. 1763.
Balsamita major Desf. Act. Soc. Hist. Nat. Paris 1:3. 1792.
Chrysanthemum balsamita (L.) var. *tanacetoides* Boiss. ex W. Miller in Bailey, Cyclop. Am. Hort. 313. 1900.

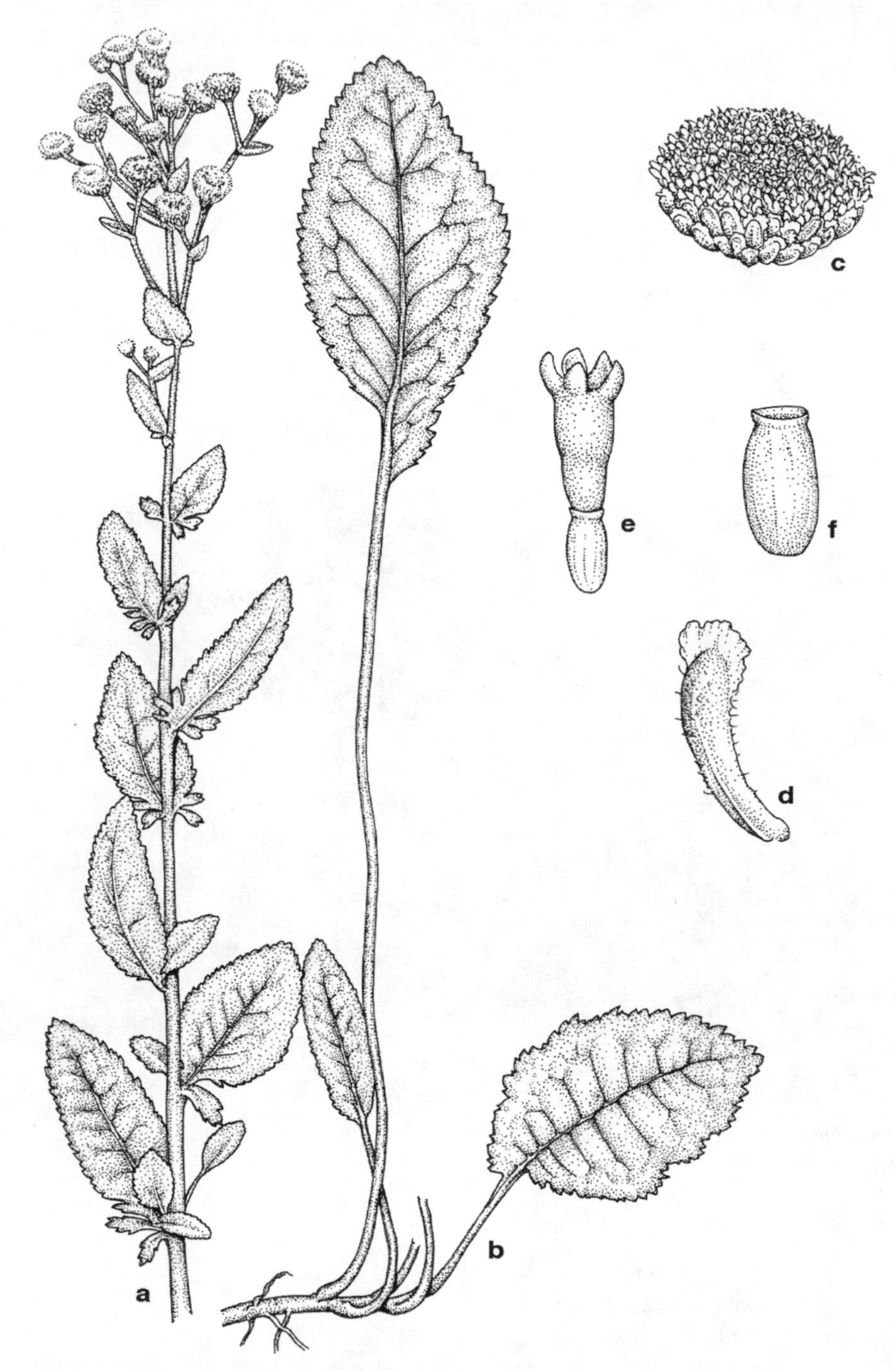

107. *Tanacetum balsamita* (Costmary).

a. Upper part of plant.
b. Basal leaves.
c. Flowering head.
d. Phyllary.
e. Disc flower.
f. Cypsela.

Perennial herb; stems erect, branched, sparsely pubescent to glabrate, to 80 cm tall; leaves basal and cauline, alternate, elliptic to oblong, often with a pair of small lobes at the base, crenate to dentate, usually pubescent, the lower ones petiolate, the middle and upper ones sessile; heads up to 60, 1.0–1.6 cm across, with or without rays, in corymbs; involucre depressed; phyllaries in 3–5 series, unequal, linear, obtuse at the apex, pubescent; receptacle epaleate; ray flowers absent or up to 15, white, up to 6 mm long, pistillate, fertile; disc flowers numerous, tubular, bisexual, yellow, 1.5–2.0 high; cypselae columnar, up to 8-ribbed, glabrous, 1.5–2.0 mm long; pappus a short crown of scales up to 0.4 mm long.

Common Name: Costmary; mint geranium.
Habitat: Disturbed soil.
Range: Native to Asia; escaped from gardens in much of the United States and Canada.
Illinois Distribution: Scattered in Illinois.

In the past, this garden escape has been known as *Chrysanthemum balsamita* or *Balsamita major*.

Sometimes the flowering heads do not produce ray flowers.

Tanacetum balsamita flowers during August and September.

22. **Achillea** L.—Milfoil; Yarrow

Perennial herbs; stems erect, branched; leaves basal and cauline; basal leaves withered at flowering time, petiolate; cauline leaves alternate, usually deeply lobed, sessile, sometimes clasping; heads several to numerous, radiate, in corymbs; involucre hemispheric to campanulate to cylindric; phyllaries in 2–4 series, unequal, scarious along the margins; receptacle flat, paleate; ray flowers up to 8, pistillate, the rays more or less orbicular, often white; disc flowers up to 75, tubular, gray, 5-lobed, bisexual; cypselae flattened, 2-ribbed, glabrous; pappus absent.

There are approximately 115 species of *Achillea* in North America, Europe, and Asia. Only the following has been found in Illinois, although *A. ptarmica* is expected, since it is known from Indiana and Missouri.

1. **Achillea millefolium** L. Sp. Pl. 2:899. 1753.

Perennial herb from short rhizomes; stems erect, branched above, pubescent to nearly glabrous, to 75 cm tall; leaves alternate, finely and deeply dissected, the lowest ones petiolate, the middle and upper ones sessile, pubescent to nearly glabrous, to 35 cm long, to 4 cm wide; heads numerous, radiate, 4–6 mm across, in corymbs; involucre cylindric, 3–4 mm high; phyllaries in 3 series, more or less equal, lanceolate to ovate, pubescent; receptacle convex, paleate, the paleae up to 4 mm long; ray flowers up to 8 in number, 3–4 mm long, suborbicular, pistillate, white or rarely pink or rose; disc flowers up to 20, tubular, gray, bisexual, 2–4 mm high; cypselae flat, broadly winged, 1–2 mm long; pappus absent.

Two varieties occur in Illinois:

a. Leaves and stems arachnoid to nearly glabrous; corymbs flat-topped . . . . 1a. *A. millefolium* var. *millefolium*

a. Leaves and stems densely woolly; corymbs round-topped . . . . 1b. *A. millefolium* var. *lanulosum*

1a. **Achillea millefolium** L. var. **millefolium**.

Achillea millefolium L. f. *roseum* E. L. Rand & Redfield, Fl. Mt. Desert Island 18. 1891.

Leaves and stems arachnoid to nearly glabrous; inflorescence flat-topped.

Common Name: Yarrow; milfoil.
Habitat: Disturbed soil; old fields; woods.
Range: Throughout North America; Mexico; Europe; Asia.
Illinois Distribution: Common; in every county.

108. *Achillea millefolium* var. *lanulosum* (Milfoil; yarrow).
a. Upper part of plant.
b. Leaf.
c. Flowering head.
d. Phyllaries.
e. Ray flower.
f. Disc flower.

Apparently some of our plants may be native while others may be adventive.

Variety *millefolium* has moderately pubescent to nearly glabrous leaves and stems and a flat-topped inflorescence.

This variety is also a garden ornamental. Rose, pink, and yellow rays are popular ornamental variants. Pink- and rose-flowered plants are rarely encountered in the wild in Illinois.

Flowering heads may be found from May through October.

1b. **Achillea millefolium** L. var. **lanulosum** (Nutt.) Piper, Mazama 2:97. 1901. Fig. 108.

Achillea lanulosum Nutt. Journ. Acad. Nat. Sci. Phila. 7:36. 1834.

Leaves and stems densely woolly; inflorescence round-topped.

Common Name: Woolly yarrow; woolly milfoil.
Habitat: Disturbed soil, old fields, woods.
Range: Throughout North America.
Illinois Distribution: Scattered throughout the state.

This plant appears rather distinct from var. *millefolium* because of its densely woolly leaves and stems and the corymb, which is round-topped. Nuttall described it as a different species. I am reluctantly considering it to be a variation of the typical variety. Some botanists do not even recognize it as a variety.

The flowering heads bloom from June to October.

23. **Chamaemelum** Mill.—Chamomile

Aromatic annuals or perennials; stems decumbent to erect; leaves alternate, pinnatifid; heads usually radiate, borne singly or a few in corymbs; involucre hemispheric; phyllaries in 3–5 series, unequal, the margins scarious; receptacle paleate; ray flowers up to 20 in number, white, rarely absent, pistillate, fertile or sterile; disc flowers numerous, bisexual, tubular, yellow, 5-lobed; cypselae oblongoid to obovoid, weakly ribbed, glabrous; pappus absent.

Two species comprise this genus, native to Europe and North Africa.

Only the following species occurs in Illinois.

1. **Chamaemelum nobile** (L.) Allioni, Fl. Pedem. 1:185. 1785. Fig. 109.

Anthemis nobilis L. Sp. Pl. 2:894. 1753.

Strongly aromatic perennial herbs; basal shoots creeping; stems decumbent to erect, much branched, softly pubescent, to 25 cm tall; leaves alternate, deeply pinnatifid, the divisions filiform to linear, softly pubescent, sessile; heads usually radiate, borne singly or few in a corymb, to 2 cm across; involucre hemispheric, 4–8 mm high; phyllaries in 3–5 series, obovate, obtuse at the apex, pubescent, with scarious margins; receptacle paleate, with obtuse paleae; ray flowers up to 20 in number, 7–10 mm long, white, pistillate; disc flowers numerous, tubular, bisexual, yellow, 2–3 mm high; cypselae oblongoid, usually 3-angled, few-ribbed, glabrous, 1.0–1.5 mm long; pappus absent.

Common Name: White chamomile; low chamomile.
Habitat: Disturbed soil.
Range: Native to Europe; sparingly introduced but rarely persisting in the United States.
Illinois Distribution: Scattered in Illinois.

This species is often mat-forming. The narrow, pinnatifid leaves are distinctive.

Chamaemelum nobile traditionally has been placed in the genus *Anthemis*, but differs by its 3-angled, fewer ribbed cypselae.

This species flowers from July to September.

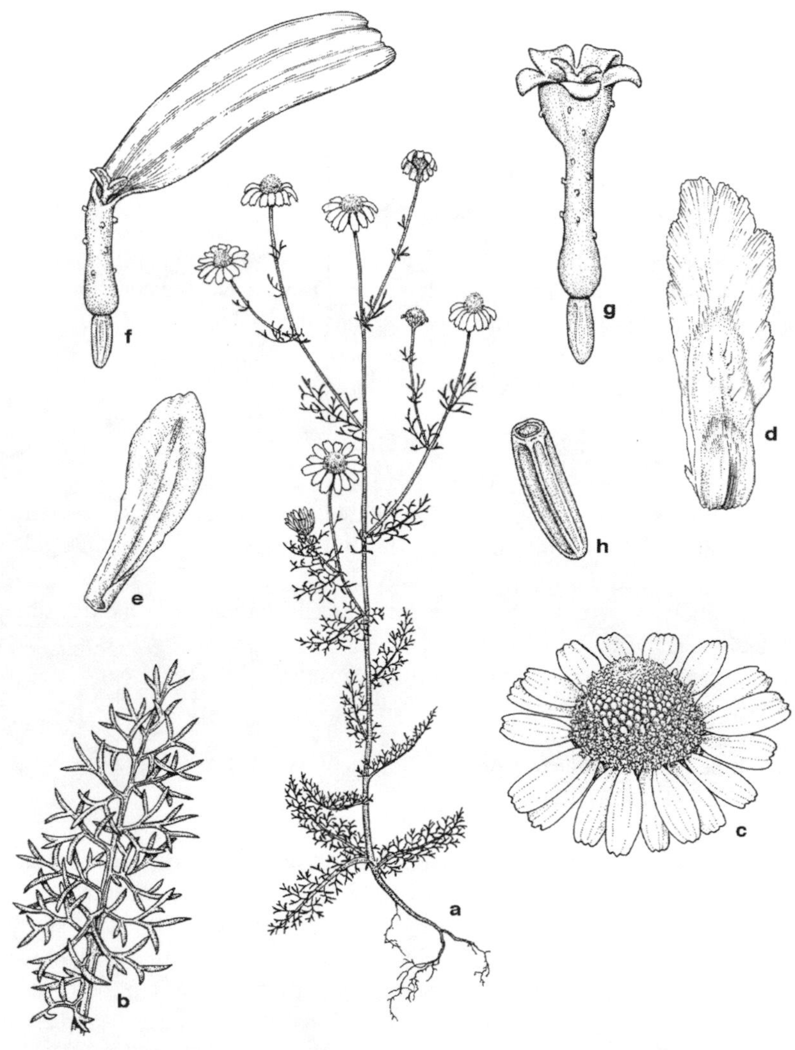

109. *Chamaemelum nobile* (White chamomile).
a. Habit.
b. Leaf.
c. Flowering head.
d, e. Phyllaries.
f. Ray flower.
g. Disc flower.
h. Cypsela.

24. **Artemisia** L.—Wormwood

Aromatic herbs or shrubs; leaves basal and also usually cauline and alternate, usually pinnately lobed or toothed or sometimes entire; heads discoid, in spikes or panicles; involucre various; phyllaries in 4–7 series, unequal, the margins usually scarious; receptacle flat, convex, or concave, paleate; central flowers of disc usually bisexual; marginal flowers of disc pistillate and fertile, or all bisexual, tubular, 2- to 5-lobed, often gray; cypselae fusiform, with 0–5 ribs, often gland-dotted; pappus absent.

There may be as many as 500 species in this genus in most parts of the world, particularly in the western United States.

1. Leaves white-tomentose or densely pubescent, at least on the lower surface.
    2. Receptacle woolly.
        3. Divisions of leaves oblong to lanceolate, 1.5–5.0 mm wide....... 1. *A. absinthium*
        3. Divisions of leaves linear to filiform, 1 mm wide or less.............. 2. *A. frigida*
    2. Receptacle glabrous.
        4. Subshrubby perennials.. 3. *A. pontica*
        4. Annual or biennial herbs.
            5. Leaves entire, serrate, or irregularly few-toothed.
                6. Leaves regularly serrate.................................... 4. *A. serrata*
                6. Leaves entire or irregularly few-toothed.................. 5. *A. ludoviciana*
            5. Leaves pinnatifid.
                7. Leaves green and glabrous above, densely white-tomentose beneath....6. *A. vulgaris*
                7. Leaves pubescent beneath but not densely white-tomentose ....7. *A. forwoodii*
1. Leaves glabrous or nearly so.
    8. Leaves or their divisions linear to filiform, entire or irregularly lobed, the lower 3- or 5-lobed or pinnatifid but not sharply toothed.
        9. Phyllaries glabrous.
            10. Leaves mostly entire, occasionally irregularly lobed or the lower 3- or 5-lobed................................... 8. *A. dracunculus*
            10. Leaves mostly pinnatifid.
                11. Stems glabrous ....9. *A. caudata*
                11. Stems gray-pubescent .... 7. *A. forwoodii*
        9. Phyllaries pubescent.. 10. *A. abrotanum*
    8. Leaves or their divisions lanceolate, sharply toothed.
        12. Inflorescence dense, spicate................................ 11. *A. biennis*
        12. Inflorescence loose, paniculate .... 12. *A. annua*

1. **Artemisia absinthium** L. Sp. Pl. 2:848. 1753. Fig. 110.

Aromatic perennial with a woody base and much branched rhizome; stems branched, gray-green, canescent, to 1 m tall; leaves alternate, gray-green, densely canescent, the lower on long petioles, 2- to 3-pinnatifid, the middle leaves pinnately lobed, the divisions oblong to linear-oblong, the upper leaves sessile, entire; heads numerous in panicles, pendulous, 3–4 mm across, on short peduncles; involucre ovoid, 2–3 mm high; phyllaries in 4–7 series, the outer linear, the inner oblong, all with scarious margins, densely sericeous; receptacle paleate, pubescent; marginal flowers 10–20 in number, pistillate; central flowers up to 50, bisexual; all flowers 1–2 mm high, glandular; cypselae obovoid, faintly ribbed, glabrous, shiny, 0.4–0.7 mm long; pappus absent.

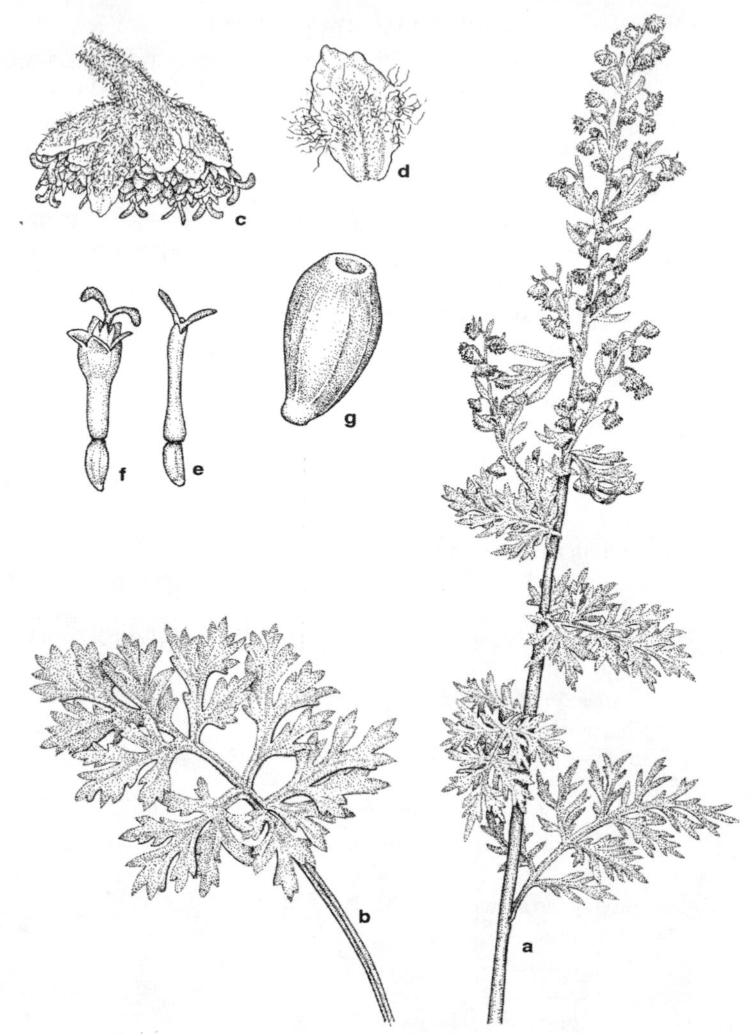

110. *Artemisia absinthium* (Absinth; common wormwood).

a. Upper part of plant.
b. Leaf.
c. Flowering head.
d. Phyllary.
e. Marginal flower.
f. Central flower.
g. Cypsela.

Common Name: Common wormwood; absinthe.
Habitat: Disturbed soil.
Range: Native to Europe; adventive in much of the United States.
Illinois Distribution: Scattered in the state.

This species is grown as an ornamental because of its attractively divided gray-green foliage. It is strongly aromatic, containing a substance known as absinthe.

Artemisia absinthium flowers from July through September.

2. **Artemisia frigida** Willd. Sp. Pl. 3:1838. 1807. Fig. 111.

Strongly aromatic perennial herb but usually with a woody base and a much-branched rhizome; stems much-branched from the base, gray-green, densely silky-canescent, to 1 m tall; leaves alternate, usually deeply 3- or 5-parted, or the upper ones undivided, each division linear and 0.5–1.0 mm wide, densely silvery-sericeous, the lowest petiolate, the upper sessile; heads discoid, numerous in racemes or panicles, nodding, 3–4 mm across, on short peduncles; involucre globose, 4–5 mm high; phyllaries in 4–7 series, gray-green, the outer linear, the inner broader, both scarious along the margins, densely tomentose; receptacle with long white hairs; marginal flowers up to 15 (–17), pistillate; central flowers up to 50, bisexual; corolla of both 1.5–2.0 mm high, glabrous; cypselae obovoid, faintly ribbed, glabrous, shiny, 1.0–1.5 mm long; pappus absent.

Common Name: Prairie sagewort; fringed sage.
Habitat: Along a railroad (in Illinois).
Range: Mostly in the western United States; adventive in Illinois.
Illinois Distribution: Known only from Cook County.

This gray-green species is native to the west of Illinois. It differs by its very narrow leaf divisions.

The flowers appear from July to September.

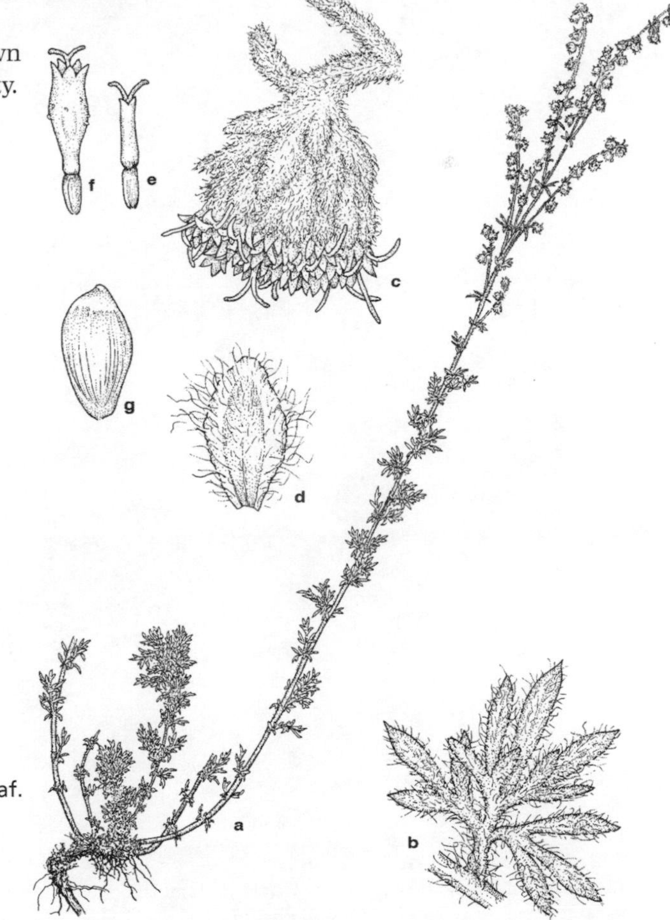

111. *Artemisia frigida* (Prairie sagewort).
a. Habit.
b. Ultimate segment of leaf.
c. Flowering head.
d. Phyllary.
e. Marginal flower.
f. Central flower.
g. Cypsela.

3. **Artemisia pontica** L. Sp. Pl. 2:847. 1753. Fig. 112.

Aromatic perennial or subshrub with creeping rhizomes; stems somewhat woody, erect, usually unbranched, brown, canescent or sometimes glabrous, to 1 m tall; leaves alternate, gray-green, long-petiolate, the lower 2- to 3-pinnate, the divisions linear, serrate, the uppermost linear, entire, sessile; heads discoid, numerous, pendulous, 3–4 mm across, in panicles; involucre hemispheric, 3–4 mm high, canescent; phyllaries in 4–7 series, linear, densely pubescent; receptacle glabrous; marginal flowers 10–12, pistillate; central flowers up to 50, bisexual, pale yellow, up to 0.3 mm high; cypselae ellipsoid, glabrous, 0.1–0.2 mm long; pappus absent.

Common Name: Roman wormwood.

Habitat: Disturbed areas.

Range: Native to Europe and Asia; introduced into the northern United States and Canada.

Illinois Distribution: Known from Cass, Lake, Morgan, and Winnebago counties.

The gray-green leaves account for the ornamental value of this species.

Artemisia pontica flowers from July to October.

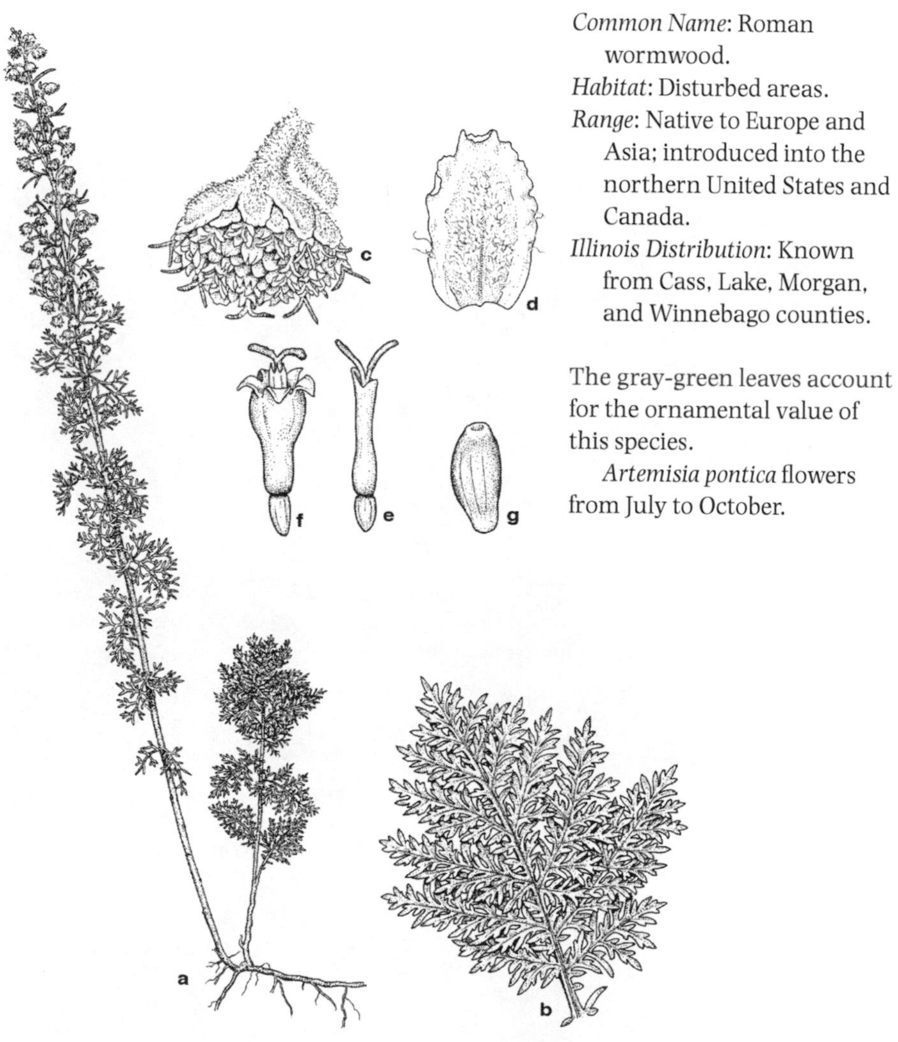

112. *Artemisia pontica* (Roman wormwood).
a. Habit.
b. Leaf.
c. Flowering head.
d. Phyllary.
e. Marginal flower.
f. Central flower.
g. Cypsela.

4. **Artemisia serrata** Nutt. Gen. N. Am. Plants 2:142. 1818. Fig. 113.
Artemisia vulgaris L. var. *serrata* (Nutt.) Torr. & Gray, Fl. N. Am. 2:420. 1843.
Artemisia vulgaris L. ssp. *serrata* (Nutt.) H. M. Hall & Clements, Publ. Carnegie Inst. Wash. 326:79. 1923.

Aromatic perennial herbs from stout rhizomes; stems erect, much branched, brown, tomentose or sometimes glabrous, to 1.2 m tall; leaves alternate, lanceolate, acute to acuminate at the apex, tapering to the sessile base, dark green above, white-tomentose below, serrate or rarely entire, to 15 cm long, to 2.5 cm wide;

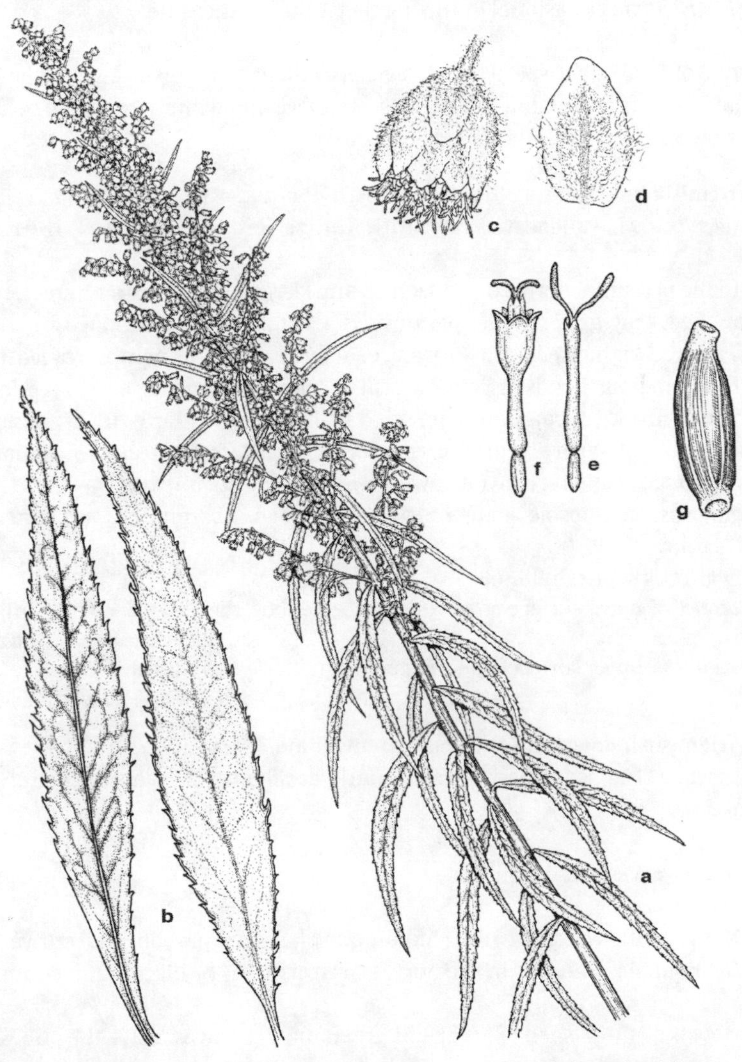

113. *Artemisia serrata* (Serrate-leaved sage).
a. Upper part of plant.
b. Leaves.
c. Flowering head.
d. Phyllary.
e. Marginal flower.
f. Central flower.
g. Cypsela.

heads discoid, numerous, green, 3–4 mm across, in racemes; involucre campanulate, 2.5–3.1 mm high; phyllaries in 4–7 series, unequal, lanceolate, obtuse to acute at the apex, densely tomentose; receptacle glabrous; marginal flowers 3–5, pistillate; central flowers about 10, bisexual, pale yellow, 1.5–2.0 mm high; cypselae ellipsoid, glabrous, about 1 mm long; pappus absent.

Common Name: Serrate-leaved sage; serrate-leaved mugwort.
Habitat: Moist ground.
Range: New York to North Dakota, south to Illinois.
Illinois Distribution: Occasional in the northern half of the state.

This is one of four native species of *Artemisia* in Illinois. It is distinguished by its lanceolate, serrate leaves that are dark green above and white-tomentose below.
The flowers bloom in August and September.

5. **Artemisia ludoviciana** Nutt. Gen. N. Am. Plants 2:143. 1818.
Artemisia vulgaris L. var. *ludoviciana* (Nutt.) Kuntze, Rev. Gen. Pl. 1:309. 1891.

Aromatic perennial herb from rhizomes and slender stolons; stems erect, usually branched, gray-green, white-pannose, to 1 m tall; leaves alternate, shallowly lobed to entire, bright green or whitish green above, white-tomentose, at least below, to 10 cm long; heads discoid, several to many, 3–4 mm across, in panicles and racemes; involucre campanulate, up to 5 mm high; phyllaries in 4–7 series, unequal, lanceolate to ovate, gray-green, tomentose; receptacle glabrous; marginal flowers up to 12, pistillate; central flowers up to nearly 50, bisexual, yellow, 1.5–3.0 mm high; cypselae ellipsoid, sometimes faintly ribbed, glabrous, 0.4–0.5 mm long; pappus absent.
Two varieties occur in Illinois:
a. Pubescence of upper surface of leaves early deciduous, the leaves becoming bright green . . . . 5a. *A. ludoviciana* var. *ludoviciana*
a. Pubescence of upper surface of leaves persistent, the leaves whitish green . . . . 5b. *A. ludivociana* var. *gnaphalodes*

5a. **Artemisia ludoviciana** Nutt. var. **ludoviciana**. Fig. 114.
Pubescence of upper surface of leaves early deciduous, the leaves becoming bright green.

Common Name: Western mugwort.
Habitat: Disturbed soil.
Range: Native to the western United States; our plants are probably adventive.
Illinois Distribution: Scattered in the northern three-fifths of Illinois; not common.

This variety is distinguished by its bright green mature leaves, since the pubescence of the young leaves does not persist. This variety is apparently not as common in Illinois as the next one.
Flowering time is July to September.

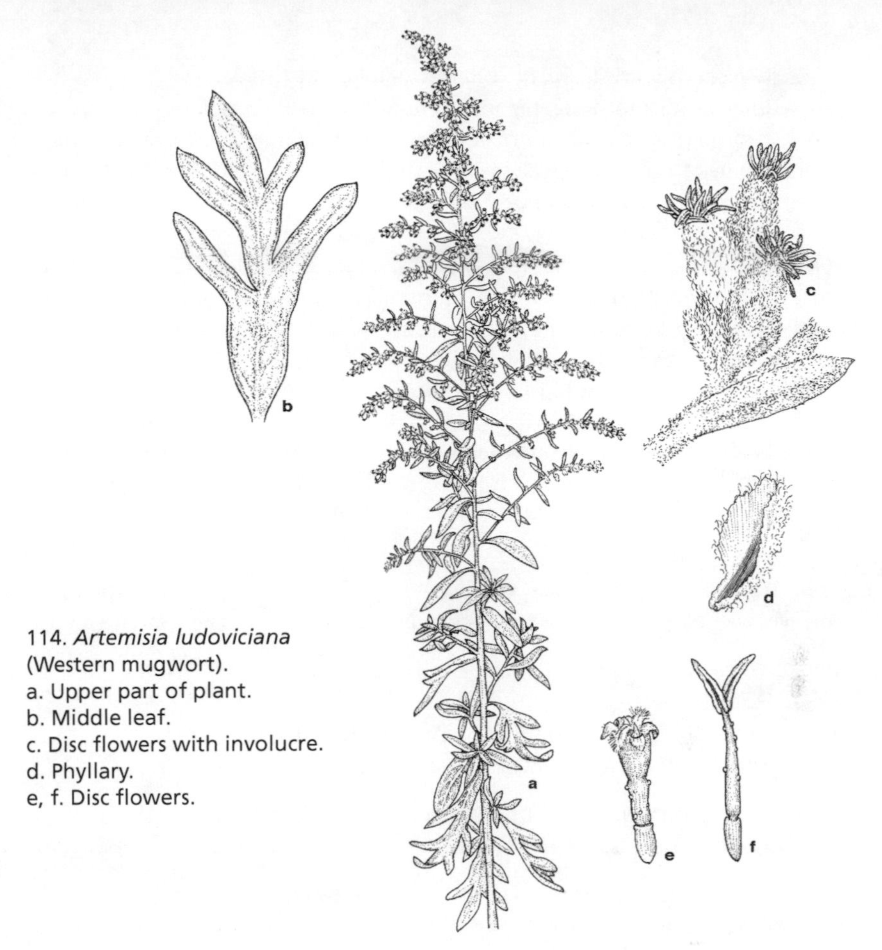

114. *Artemisia ludoviciana* (Western mugwort).
a. Upper part of plant.
b. Middle leaf.
c. Disc flowers with involucre.
d. Phyllary.
e, f. Disc flowers.

5b. **Artemisia ludoviciana** Nutt. var. **gnaphalodes** (Nutt.) Torr. & Gray, Fl. N. Am. 2:420. 1843.
Artemisia gnaphalodes Nutt. Gen. N. Am. Plants 2:143. 1818.

Pubescence of upper surface of leaves persistent, the leaves whitish green.

Common Name: White sage; prairie sage.
Habitat: Disturbed soil.
Range: Ontario to Alberta, south to Texas and Illinois; Mississippi.
Illinois Distribution: Occasional in the northern three-fifths of Illinois; also Jackson and St. Clair counties.

The whitish green leaves of this variety are distinctive. Superficially, this variety looks very different from var. *ludoviciana*, but the characteristics of the flowering heads are not significantly different from the typical variety. Some botanists in the past have recognized this as a distinct species.

The flowering heads bloom from July to September.

6. **Artemisia vulgaris** L. Sp. Pl. 2:848. 1753. Fig. 115.

Perennial herb from stout rhizomes but without stolons; stems erect, usually branched, brown or reddish, more or less glabrous, to 2 m tall; leaves alternate, variable, some of them deeply cleft, some shallowly cleft, some entire, glabrous or pubescent, green or sometimes densely white-tomentose beneath, to 10 cm long; heads discoid, numerous, erect, 3–4 mm across, in panicles or racemes; involucre ovoid, 3–4 mm high; phyllaries in 4–7 series, unequal, lanceolate, pubescent or nearly glabrous; receptacle glabrous; marginal flowers 7–10, pistillate; central flowers up to 20, bisexual, yellow or reddish, 1.5–3.0 mm high; cypselae ellipsoid, glabrous, up to 1.2 mm long; pappus absent.

Three varieties occur in Illinois:

a. Leaves cleft at least halfway to the midvein, the teeth either acuminate or absent.
  b. Lobes of leaves with acuminate teeth . . . . .6a. *A. vulgaris* var. *vulgaris*
  b. Lobes of leaves entire. . . . . 6b. *A. vulgaris* var. *glabra*
a. Leaves cleft only about one-fourth the way to the midvein, the teeth obtuse to subacute . . . . 6c. *A. vulgaris* var. *latiloba*

6a. **Artemisia vulgaris** L. var. **vulgaris**

Leaves cleft at least halfway to the midvein, the lobes with acuminate teeth.

Common Name: Common mugwort.
Habitat: Disturbed areas.
Range: Native to Europe; introduced in much of the United States.
Illinois Distribution: Scattered in the northern half of the state.

This apparently is the most common variety of *A. vulgaris* in Illinois.

The flowering time is July to October.

6b. **Artemisia vulgaris** L. var. **glabra** Ledeb. Fl. Altaic. 4:83. 1833.

Leaves cleft at least halfway to the midvein, the lobes entire.

Common Name: Common mugwort.
Habitat: Disturbed soil.
Range: Native to Europe; introduced in the United States.
Illinois Distribution: Scattered in disturbed areas, mostly in the northern half of the state.

This variety flowers from July to October.

6c. **Artemisia vulgaris** L. var. **latiloba** Ledeb. Fl. Altaic. 4:83. 1833.

Leaves cleft about one-fourth of the way to the midvein, the lobes obtuse to subacute.

Common Name: Common mugwort.
Habitat: Disturbed areas.
Range: Native to Europe; introduced in the United States.
Illinois Distribution: Scattered in northern Illinois.

Because of the less deeply cleft leaves, this variety looks quite a bit different from the other two varieties in the state.

The flowers appear from July to October.

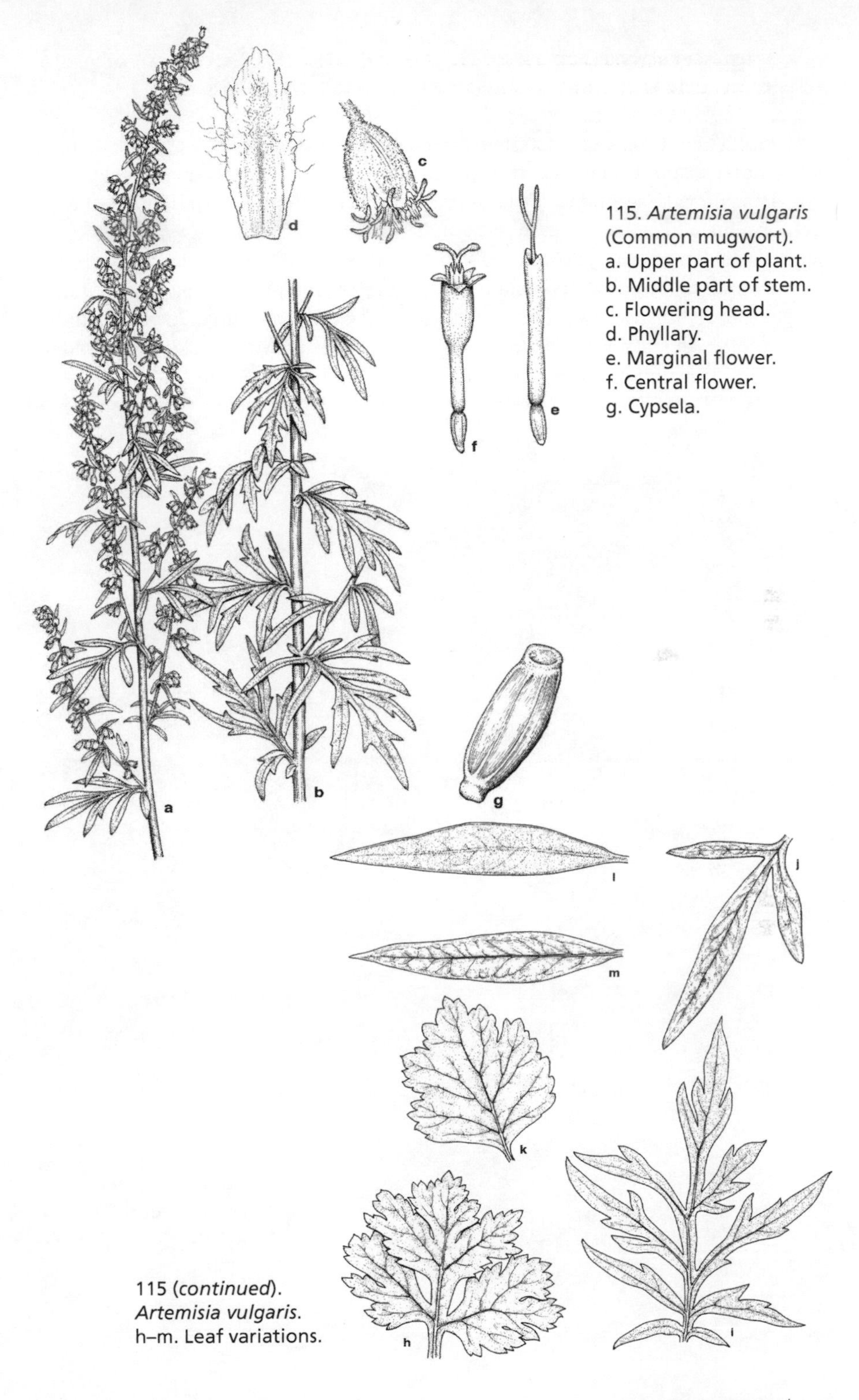

115. *Artemisia vulgaris* (Common mugwort).
a. Upper part of plant.
b. Middle part of stem.
c. Flowering head.
d. Phyllary.
e. Marginal flower.
f. Central flower.
g. Cypsela.

115 (*continued*).
Artemisia vulgaris.
h–m. Leaf variations.

7. **Artemisia forwoodii** S. Wats. Proc. Am. Acad. Arts & Sci. 25:133. 1890. Fig. 116.
Artemisia caudata Michx. var. *calvens* Lunell, Am. Midl. Nat. 2:188. 1912.

Biennial herb from a taproot; stems erect, canescent-pilose, to 1.5 m tall; first year's leaves all basal, petiolate, gray-pubescent, 2- to 3-pinnately divided, to 12 cm long, to 5 cm wide; second year's leaves cauline, alternate, 2- to 3-pinnatifid, the divisions linear, acute at the apex, gray-pubescent, to 4 cm long, to 1.5 cm wide; heads discoid, numerous, crowded, 2–3 mm across, in panicles or racemes; involucre turbinate, 2–3 mm high; phyllaries in 4–7 series, unequal, oblong, pubescent but glabrate, with scarious margins; receptacle glabrous; marginal flowers up to 20, pistillate; central flowers up to 30, staminate, pale yellow; cypselae oblongoid to lanceoloid, indistinctly nerved, glabrous, 0.8–1.0 mm long; pappus absent.

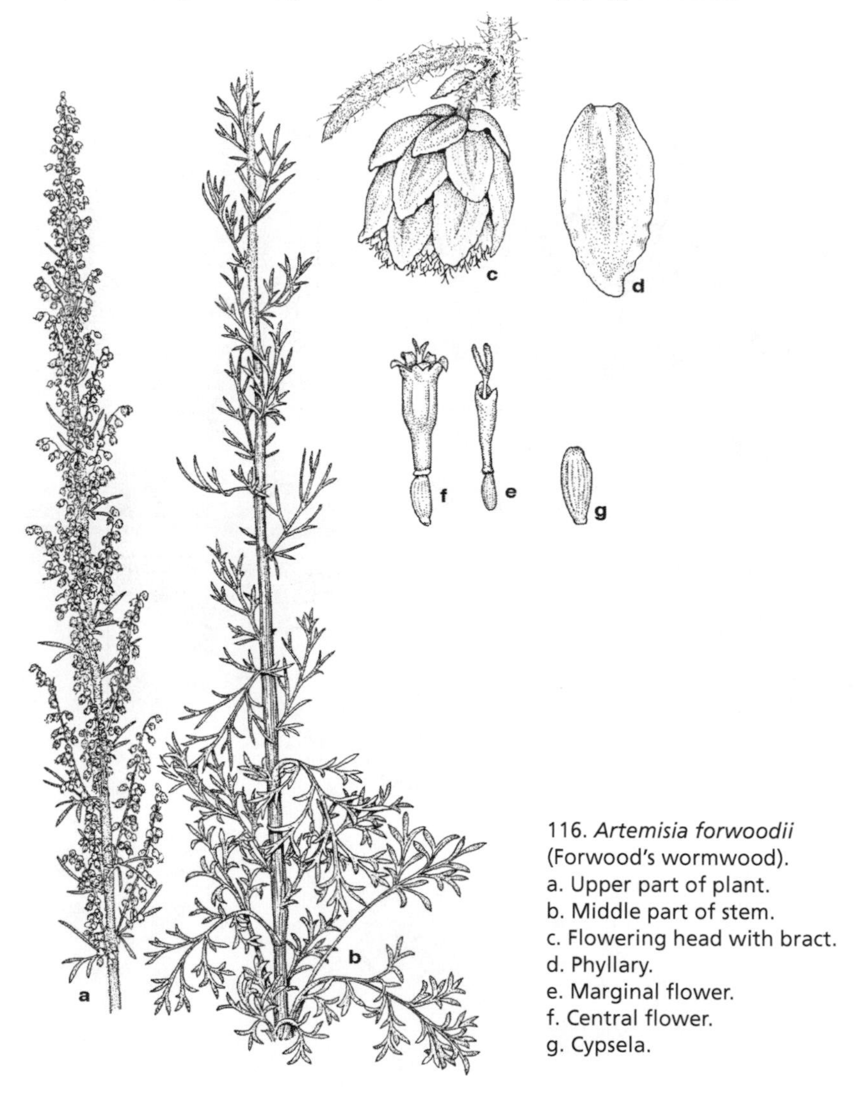

116. *Artemisia forwoodii* (Forwood's wormwood).
a. Upper part of plant.
b. Middle part of stem.
c. Flowering head with bract.
d. Phyllary.
e. Marginal flower.
f. Central flower.
g. Cypsela.

Common Name: Forwood's wormwood; gray beach wormwood.
Habitat: Sand dunes.
Range: Quebec to Manitoba, south to Texas and Florida.
Illinois Distribution: Cook and Lake counties.

Although some botanists include *A. forwoodii* within *A. campestris*, others include it within *A. caudata*. The gray pubescence throughout the plant is distinctive and readily distinguishes *A. forwoodii* from either *A. campestris* or *A. caudata*. I consider it to be a distinct species.

Artemisia forwoodii grows on sand dunes near Lake Michigan.

Flowering time is August to October.

8. **Artemisia caudata** Michx. Fl. Bor. Am. 2:129. 1803. Fig. 117.
Artemisia campestris L. ssp. *caudata* (Michx.) H. M. Hall & Clements, Publ. Carnegie Inst. Wash. 326:122. 1923.

117. *Artemisia caudata* (Sand wormwood).
a. Upper part of plant.
b. Middle part of stem.
c. Leaf.
d. Flowering head.
e. Phyllary.
f. Marginal flower.
g. Central flower.
h. Cypsela.

Biennial herb from a taproot; stems erect, green, glabrous or nearly so, to 1.5 m tall; first year's leaves all basal, petiolate, green, glabrous or nearly so, 2- to 3-pinnately divided, to 12 cm long, to 5 cm wide; second year's leaves cauline, alternate, 2- to 3-pinnatifid, the lobes linear, acute at the apex, green, glabrous or nearly so, to 4 cm long, to 1.5 cm wide; heads discoid, numerous, crowded, 2–3 mm across, in panicles or racemes; involucre turbinate, 2–3 mm high; phyllaries in 4–7 series, unequal, oblong to ovate, glabrous or nearly so, with scarious margins; receptacle glabrous; marginal flowers up to 20, pistillate; central flowers up to 30, staminate, pale yellow; cypselae broadly lanceoloid, faintly nerved, glabrous, 0.8–1.0 mm long; pappus absent.

Common Name: Sand wormwood; beach wormwood.
Habitat: Sandy soil.
Range: Ontario to Manitoba, south to New Mexico and Florida.
Illinois Distribution: Occasional in the northern half of Illinois; also Jersey County.

Although similar in many characteristics to *A. forwoodii*, *Artemisia caudata* differs by its green and usually glabrous stems, leaves, and phyllaries.
This species flowers from July to October.

9. **Artemisia dracunculus** L. Sp. Pl. 2:849. 1753. Fig. 118.
Artemisia cernua Nutt. Gen. 2:143. 1818.
Artemisia glauca Pallas ex Willd. Sp. Pl., ed. 4, 3:1831–1832. 1803.
Artemisia dracunculoides Pursh, Fl. Am. Sept. 2:742. 1814.
Artemisia dracunculina S. Wats. Proc. Am. Acad. Arts & Sci. 23:279. 1888.
Artemisia glauca Pallas ex Willd. var. *cernua* (Nutt.) Bush, Am. Midl. Nat. 11:27. 1928.
Artemisia dracunculoides Pursh var. *dracunculina* (S. Wats.) S. F. Blake, Journ. Wash. Acad. Sci. 30:472. 1940.

Perennial herb from rhizomes; stems erect, branched, green or brown, glabrous or gray-pubescent; basal leaves linear or sometimes 3-cleft, to 8 cm long, to 1.2 cm wide, green or gray-green, glabrous or somewhat pubescent; cauline leaves alternate, linear or sometimes lobed, acute the apex, to 7 cm long, to 1 cm wide, bright green, usually glabrous; heads discoid, numerous, pendulous, to 2 mm across, in panicles; involucre hemispheric to globose, 2–3 mm high; phyllaries in 4–7 series, unequal, ovate to lanceolate, acute at the apex, glabrous or nearly so, the margins scarious; receptacle glabrous; marginal flowers up to 25, pistillate; central flowers up to 20, staminate, pale yellow, 1.5–2.0 mm high; cypselae oblongoid, faintly nerved, glabrous, 0.5–1.0 mm long; pappus absent.

Common Name: Wild tarragon; false tarragon.
Habitat: Prairies.
Range: Ontario to Alaska, south to California and Texas; Europe; Asia.
Illinois Distribution: Rare in the northern half of the state.

This species is the tarragon used in cooking.

Plants with pubescent stems and leaves have been called *A. glauca*. Glabrous plants are typical *A. dracunculus*. Those plants with flowering heads on peduncles more than 3 mm long have been called *A. dracunculina*.

Artemisia dracunculus flowers from July through September.

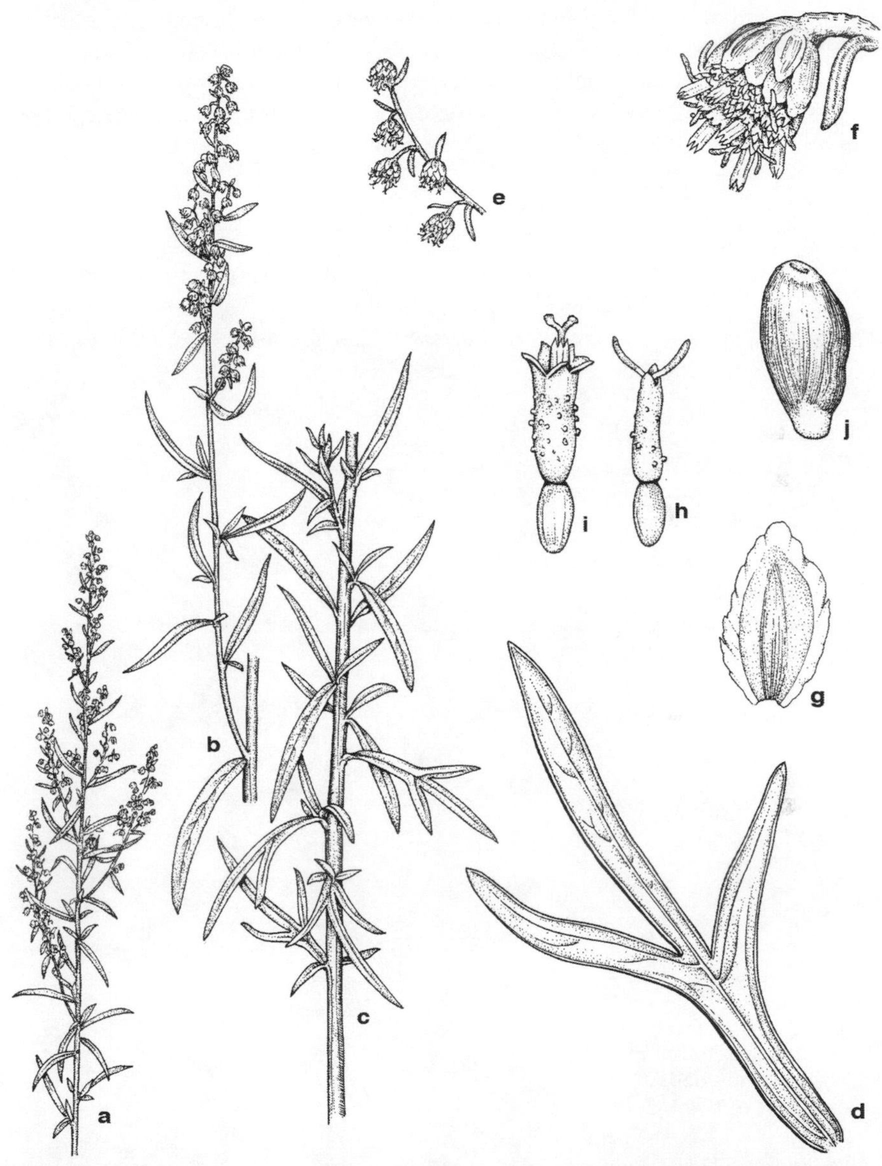

118. *Artemisia dracunculus* (Wild tarragon).
a, b. Upper part of plant.
c. Middle part of stem.
d. Ultimate leaf segment.
e. Cluster of flowering heads.
f. Flowering head.
g. Phyllary.
h. Marginal flower.
i. Central flower.
j. Cypsela.

10. **Artemisia abrotanum** L. Sp. Pl. 2:845. 1753. Fig. 119.

Perennial herb, usually with a woody stem near the base; stems erect, branched, brown, glabrous or sparsely pubescent, to 1.2 m tall; leaves alternate, simple, 2- to 3-pinnatifid, the divisions filiform to linear, entire, dark green, glabrous on the upper surface, sparsely pubescent on the lower surface; heads discoid, numerous, 2.5–3.0 mm across, nodding, in racemes or panicles; involucre hemispheric to ovoid, 1.3–3.5 mm high; phyllaries in 4–7 series, oblong to lanceolate, obtuse to acute at the apex, sparsely pubescent; receptacle glabrous; marginal flowers up to 10, pistillate; central flowers up to 20, bisexual, yellow, 0.5–1.0 mm high; cypselae ellipsoid, often angular, glabrous, 0.5–1.0 mm long; pappus absent.

Common Name: Southern wormwood.
Habitat: Disturbed areas.
Range: Native to Europe, Asia, and Africa; introduced but usually not persisting in the United States.
Illinois Distribution: Known only from Cook and LaSalle counties, where it is an escape from cultivation.

This species is distinctive by its filiform to linear divisions of the leaves and its yellow flowering heads.

Artemisia abrotanum flowers in August and September.

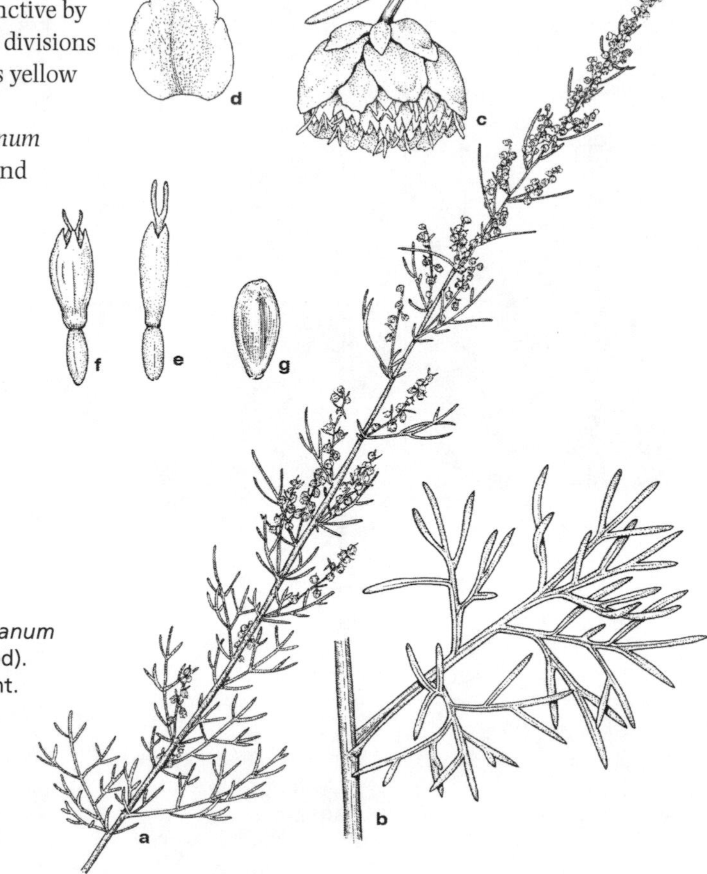

119. *Artemisia abrotanum* (Southern wormwood).
a. Upper part of plant.
b. Leaf.
c. Flowering head.
d. Phyllary.
e. Marginal flower.
f. Central flower.
g. Cypsela.

11. **Artemisia annua** L. Sp. Pl. 2:847. 1753. Fig. 120.

Strongly aromatic annual herb; stems erect, much branched, green to reddish brown, glabrous or nearly so, to 2 m tall; leaves alternate, 2- to 5-pinnate, the divisions lanceolate, sharply toothed, bright green, glabrous, gland-dotted, the uppermost usually sessile; heads discoid, numerous, pendulous, 2–3 mm across, in panicles; involucre hemispheric to globose, 1.5–2.5 mm high; phyllaries in 4–7 series, oblong to lanceolate to ovate, mostly acute at the apex, glabrous; receptacle glabrous; marginal flowers up to 20, pistillate; central flowers up to 20 (–25), bisexual, pale yellow, 0.5–1.0 mm high; cypselae flattened, oblong, glabrous, 0.5–0.8 mm long; pappus absent.

Common Name: Annual wormwood; sweet Annie.
Habitat: Disturbed soil.
Range: Native to Europe and Asia; adventive throughout the United States.
Illinois Distribution: Occasional throughout the state.

The sweet aroma of this plant is due to the presence of aromatic oils. It occurs in disturbed sites throughout the state.

Artemisia annua flowers from August to October.

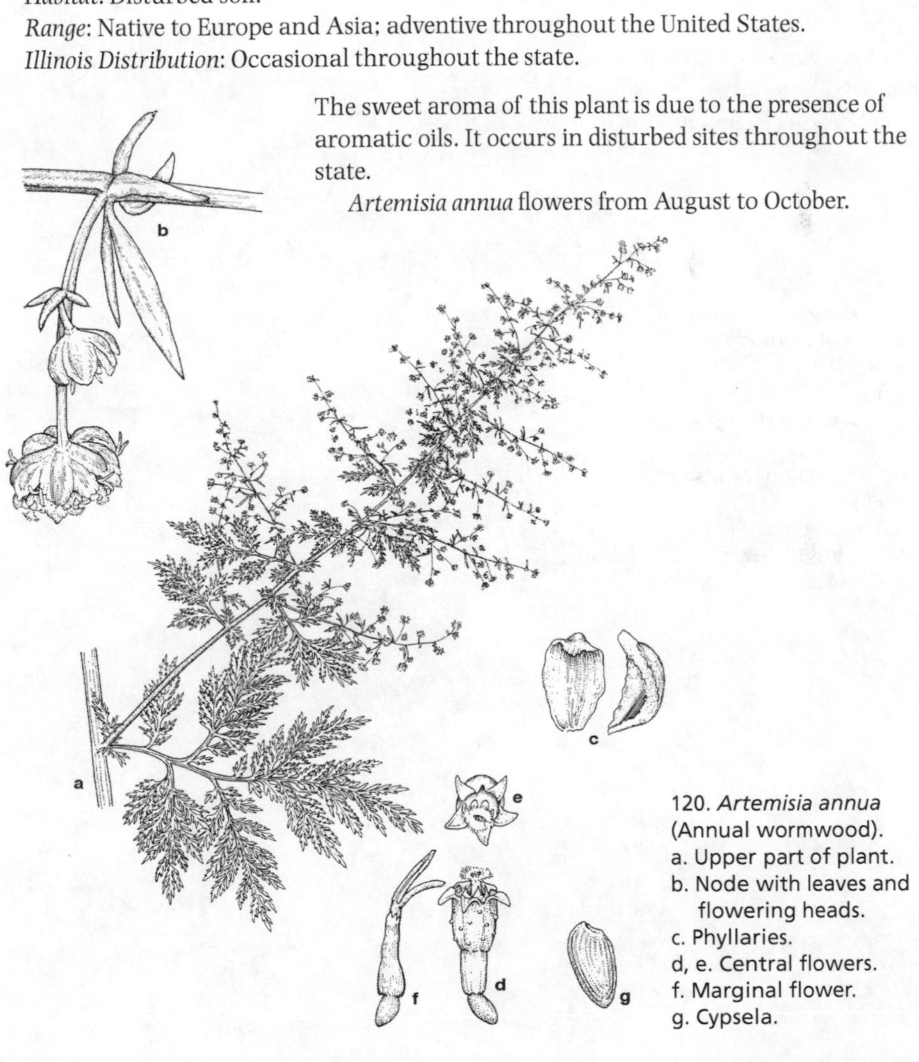

120. *Artemisia annua* (Annual wormwood).
a. Upper part of plant.
b. Node with leaves and flowering heads.
c. Phyllaries.
d, e. Central flowers.
f. Marginal flower.
g. Cypsela.

12. **Artemisia biennis** Willd. Phytographia 11. 1794. Fig. 121.

Annual or biennial herb, not aromatic; stems erect, branched, reddish, glabrous, to 1.3 m tall; leaves alternate, 1- to 2-pinnate, the divisions linear to narrowly oblong, serrate or less commonly entire, glabrous; heads discoid, numerous, 2–3 mm across, ascending to erect, in crowded, axillary spikes; involucre hemispheric, 2–3 mm high; phyllaries in 4–7 series, elliptic to obovate, green, glabrous; marginal flowers up to 25, pistillate; central flowers up to 40, bisexual, pale yellow, 1–2 mm high; cypselae ellipsoid, finely nerved, glabrous, up to 1 mm long; pappus absent.

Common Name: Biennial wormwood.
Habitat: Disturbed soil.
Range: Throughout most of the United States and Canada.
Illinois Distribution: Occasional throughout the state.

This species is not aromatic. The ascending to erect flowering heads are crowded into axillary spikes.

Artemisia biennis flowers from August to October.

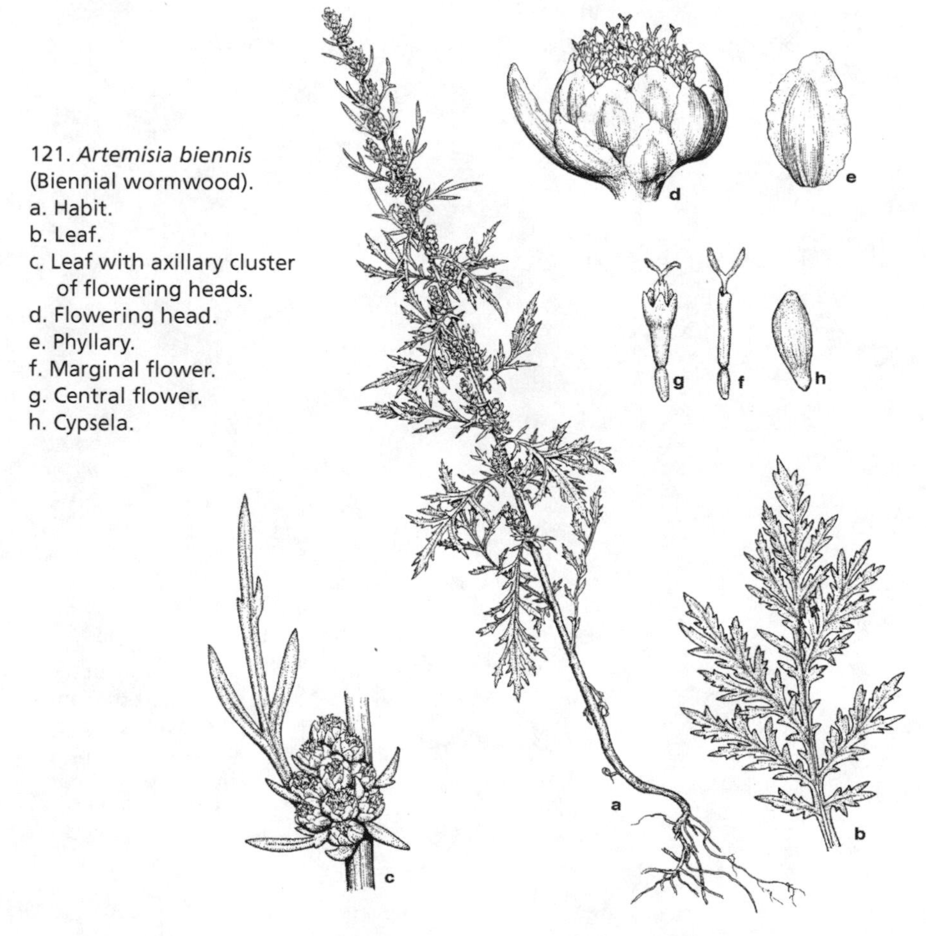

121. *Artemisia biennis* (Biennial wormwood).
a. Habit.
b. Leaf.
c. Leaf with axillary cluster of flowering heads.
d. Flowering head.
e. Phyllary.
f. Marginal flower.
g. Central flower.
h. Cypsela.

25. **Anthemis** L.—Chamomile

Annual, biennial, or perennial herbs or subshrubs, often aromatic; stems decumbent to erect; leaves alternate, sessile or petiolate, pinnately lobed; heads radiate, usually in corymbs; involucre hemispheric to obconic; phyllaries in 3–5 series, unequal, the margins scarious; receptacle convex, paleate; ray flowers pistillate or sometimes sterile, usually white; disc flowers numerous, tubular, usually yellow, bisexual, 5-lobed; cypselae terete or 4-angled in cross-section, with 9 or 10 strong ribs, glabrous; pappus a crown of minute scales or absent.

Approximately 175 species native to Europe, Asia, and Africa comprise the genus. Two occur as adventives in Illinois. A third species, previously called *Artemisia tinctoria*, is now considered to be in the genus *Cota*.

1. Achenes 4-angled, not glandular-tuberculate; plants odorless; ray flowers pistillate ..... 1. *A. arvensis*
1. Achenes subterete, glandular-tuberculate; plants with a foul odor; ray flowers sterile ..... 2. *A. cotula*

1. **Anthemis arvensis** L. Sp. Pl. 2:894. 1753. Fig. 122.
Anthemis agrestis Wallr. Sched. Crit. 484. 1822.
Anthemis arvensis L. var. *agrestis* (Wallr.) DC. Prodr. 6:6. 1837.

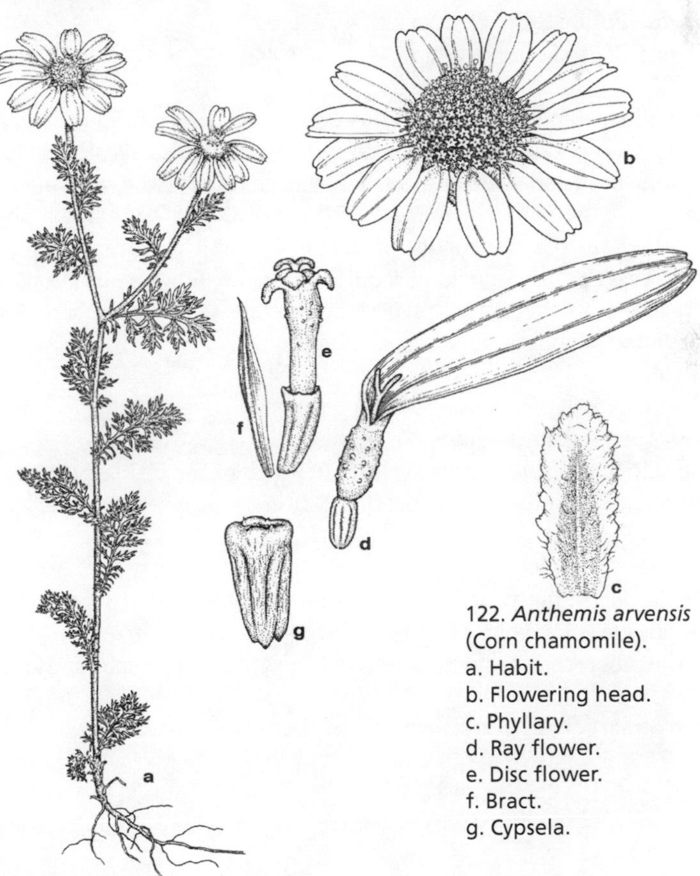

122. *Anthemis arvensis* (Corn chamomile).
a. Habit.
b. Flowering head.
c. Phyllary.
d. Ray flower.
e. Disc flower.
f. Bract.
g. Cypsela.

Annual herb without a foul odor; stems decumbent to ascending to erect, much branched, pubescent, to 75 cm tall; leaves alternate, 1- to 2-pinnate, the divisions linear to lanceolate, sessile, to 35 cm long, to 1.5 cm wide; heads radiate, few to several, to 3 cm across; involucre hemispheric to obconic, 6–10 mm high; phyllaries in 3–5 series, unequal, oblong, obtuse at the apex, pubescent; receptacle convex, paleate; ray flowers up to 20, 5–15 mm long, white, pistillate; disc flowers numerous, tubular, yellow, bisexual, 2.0–3.5 mm high; cypselae 4-angled, not glandular-tuberculate, strongly ribbed, glabrous, 1.7–2.2 mm long; pappus a crown of minute scales.

Common Name: Corn chamomile.
Habitat: Waste ground, fields, roadsides.
Range: Native to Europe; adventive in much of the United States.
Illinois Distribution: Occasional throughout the state.

This European species is found occasionally in disturbed areas in Illinois. Our plants have been called var. *agrestis*, but this variety is apparently not significant.

Anthemis arvensis flowers from May to August.

2. **Anthemis cotula** L. Sp. Pl. 2:894. 1753. Fig. 123.
Maruta cotula DC. Prodr. 6:13. 1837.

Annual foul-smelling herb from fibrous roots; stems erect, branched, glabrous or nearly so, to 85 cm tall; leaves alternate, 1- to 3-pinnate, sessile, the divisions filiform, mostly glabrous, to 5 cm long, to 3 cm wide; heads radiate, several to numerous, to 2.2 cm across; involucre hemispheric to obconic, 5–10 mm high; phyllaries in 3–5 series, oblong, obtuse, usually tomentose; receptacle convex, paleate; ray flowers up to 15 (-18), 5–15 mm long, sterile, white; disc flowers numerous, tubular, bisexual, yellow, 2.0–2.5 mm high; cypselae subterete, glandular-tuberculate, 1.5–2.0 mm long; pappus absent.

Common Name: Dog-fennel; mayweed.
Habitat: Disturbed soil, particularly barnyards.
Range: Native to Europe and Asia; adventive in much of the United States.
Illinois Distribution: Common in the northern three-fourths of Illinois, less common elsewhere.

This species has a distinctive bad odor. The subterete glandular-tuberculate cypselae and the absence of a pappus further distinguish it from *A. arvensis*.

During the nineteenth century, Illinois botanists called this species *Maruta cotula*.

Anthemis cotula flowers from May to November.

123. *Anthemis cotula* (Dog-fennel; mayweed).

a. Habit.
b. Leaves and flowering heads.
c. Leaf.
d. Flowering head.
e. Phyllary.
f. Ray flower.
g. Disc flower.
h. Cypsela.

26. **Matricaria** L.—Mayweed; Chamomile

Usually aromatic annual herbs from a taproot; leaves basal and cauline, the basal withered at anthesis, the cauline alternate, pinnately divided; heads radiate or discoid, borne singly or in corymbs; involucre hemispheric; phyllaries in 2–4 series, nearly equal, scarious along the margins; receptacle epaleate; ray flowers absent or up to 20, white, pistillate and fertile when present; disc flowers very numerous, tubular, bisexual, yellow or yellow-green, 4- or 5-lobed, bisexual, fertile; cypselae obconic, 3- to 5-ribbed, glabrous; pappus a crown of short scales or absent.

Seven species comprise this genus. Two are adventive in Illinois:

1. Heads rayless; corolla of disc flowers 4-lobed; plants with a pineapple scent when crushed . . . . 1. *M. discoidea*
1. Heads with white rays; corolla of disc flowers 5-lobed; plants without a pineapple scent when crushed . . . . 2. *M. chamomilla*

1. **Matricaria discoidea** DC. in A. DC. Prodr. 6:50. 1838. Fig. 124.

Artemisia matricarioides Less. Linnaea 6:210. 1831.

Matricaria matricarioides (Less.) Porter, Mem. Torrey Club 5:341. 1894, *nomen invalidum*.

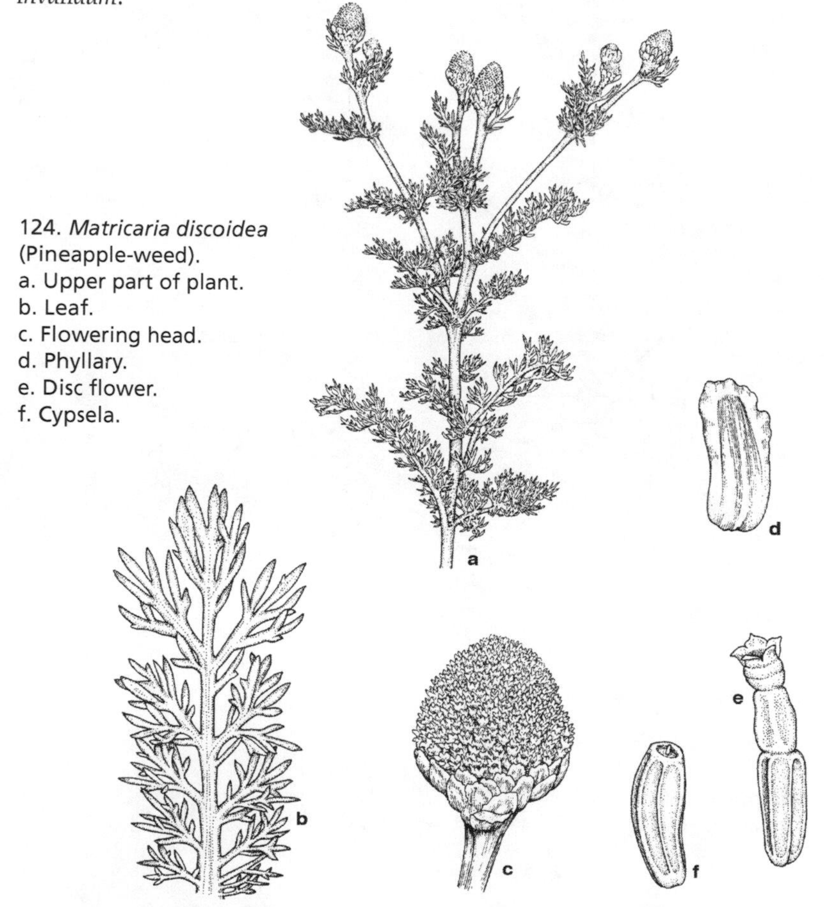

124. *Matricaria discoidea* (Pineapple-weed).
a. Upper part of plant.
b. Leaf.
c. Flowering head.
d. Phyllary.
e. Disc flower.
f. Cypsela.

Annual herb from a taproot, with a pineapple scent when crushed; stems erect or sometimes decumbent, much branched from the base, glabrous, to 45 cm tall; cauline leaves alternate, 2- to 3-pinnately divided, to 7.5 cm long, to 1.5 cm wide, the lobes linear, entire, glabrous; heads discoid, few to numerous, to 7 mm across, borne singly or in corymbs; involucre hemispheric, 2.0–3.5 mm high; phyllaries in 2 or 3 series, oblong, obtuse to acute at the apex, glabrous, with a broad, scarious margin; receptacle epaleate; ray flowers absent; disc flowers very numerous, tubular, bisexual, yellow-green, 4-lobed, 1.0–1.3 mm high; cypselae oblongoid, 3-nerved, glabrous, 1.0–1.5 mm long; pappus a crown of short scales.

Common Name: Pineapple-weed.
Habitat: Disturbed soil.
Range: Apparently native to northern and northeastern United States; adventive in Illinois.
Illinois Distribution: Common throughout the state.

This plant is readily recognized by its pineapple scent when crushed and by its small, yellow-green, rayless flowering heads.

For many years this plant was known as *Matricaria matricarioides*, but that binomial is not available for this plant.

This species flowers from April through September.

2. **Matricaria chamomilla** L. Sp. Pl. 2:891. 1753. Fig. 125.
Matricaria recutita L. Sp. Pl. 2:891. 1753.

Aromatic annual herb from a taproot; stems erect, much branched, glabrous, to 75 cm tall; leaves alternate, to 7 cm long, to 1.5 cm wide, pinnately divided into linear, entire lobes, glabrous; heads radiate, numerous, to 2 cm across; involucre hemispheric, 2–3 mm high; phyllaries in 3 series, oblong, obtuse at the apex; receptacle ovoid or conical, epaleate; ray flowers up to 20 (-25), up to 9 mm long, white, pistillate, fertile; disc flowers very numerous, tubular, yellow to greenish yellow, 5-lobed, 1.6–1.8 mm high, bisexual; cypselae oblongoid to obovoid, tan, glabrous, with 5 faint ribs, glabrous, glandular, up to 1 mm long; pappus absent or rarely with a crown of minute scales.

Common Name: German chamomile.
Habitat: Disturbed soil.
Range: Native to Europe and Asia; introduced in much of the United States and Canada.
Illinois Distribution: Occasional in the southern half of Illinois, north to Cook and Woodford counties.

In the past, this plant has been known as *Matricaria recutita*. It has less divided leaves than *M. discoidea*, conspicuous white ray flowers, and lacks the pineapple scent, although it is aromatic.

This species flowers from May to August.

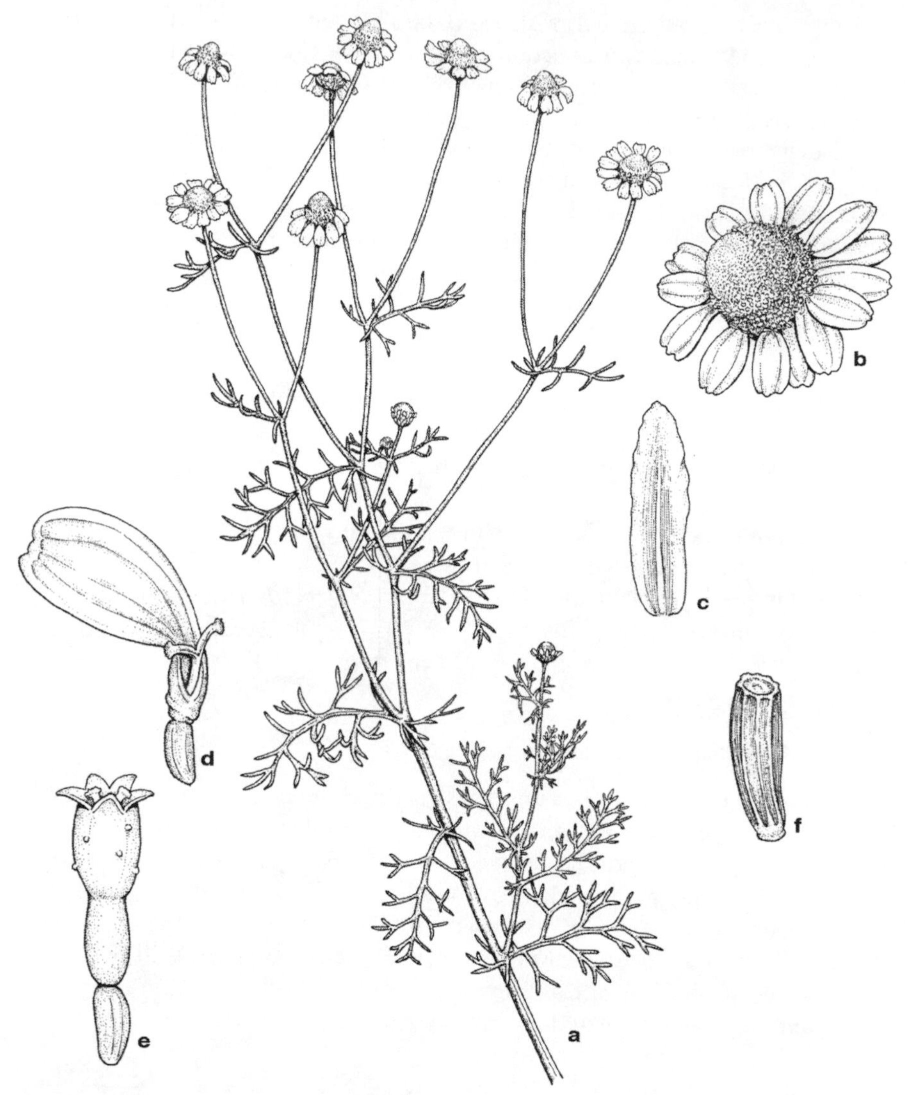

125. *Matricaria chamomilla* (German chamomile). a. Upper part of plant. b. Flowering head. c. Phyllary. d. Ray flower. e. Disc flower. f. Cypsela.

27. **Cota** J. Gay & Gussone—Yellow Chamomile

Annual, biennial, or perennial herbs, sometimes rhizomatous; stems usually pubescent; leaves alternate, pinnately divided; heads radiate, in corymbs or borne singly; involucre usually hemispheric; phyllaries in 3–5 series, unequal, with scarious margins; receptacle paleate; ray flowers yellow, pistillate or sterile; disc flowers numerous, tuberous, yellow, bisexual, 5-lobed; cypselae somewhat flattened or 4-angled, without strong ribs, glabrous; pappus a crown of minute scales.

About 40 species, all native to Europe, Asia, and Africa, comprise this genus. One is adventive in the United States.

Cota differs from *Anthemis* by its less distinct ribs of its cypselae and its paleate receptacle. Our plant used to be classified in the genus *Anthemis*.

1. **Cota tinctoria** (L.) J. Gay & Gussone, Fl. Sicul. Syn. 2:867. 1845. Fig. 126.
Anthemis tinctoria L. Sp. Pl. 2:896. 1753.

Perennial herb, usually rhizomatous; stems erect, branched, pubescent, to 1 m tall; leaves alternate, pinnately divided, the divisions linear to oblong, sharply serrate, to 3.5 cm long, to 2.2 cm wide, pubescent; heads radiate, few, to 2 cm across, on long peduncles; involucre hemispheric, 4–6 mm high; phyllaries in 3–5 series, unequal, oblong, obtuse at the apex, densely tomentose; receptacle convex, with paleae 4–5 mm long; ray flowers up to 30, up to 12 mm long, yellow, notched at the tip,

126. *Cota tinctoria* (Yellow chamomile).
a. Upper part of plant.
b. Leaf.
c. Ultimate segment of leaf.
d. Flowering head.
e. Phyllary.
f. Disc flower.
g. Bract.
h. Ray flower.
i. Cypsela.

pistillate or sometimes sterile; disc flowers numerous, tubular, bisexual, yellow, 3.5–4.0 mm high, 5-lobed, perfect; cypselae 4-angled, not strongly ribbed, 1.8–2.2 mm long; pappus with scales up to 2 mm long.

Common Name: Yellow chamomile.
Habitat: Disturbed areas.
Range: Native to Europe and Asia; scattered as garden escapes in the United States.
Illinois Distribution: Cook, DuPage, Jo Daviess, Kendall, and Winnebago counties.

Cota tinctoria usually has been included in the genus *Anthemis*, but the yellow rays and 4-angled cypselae may justify its separation into the genus *Cota*.

This species flowers from June to September.

28. **Tripleurospermum** Schultz-Bip.—Mayweed

Annual, biennial, or perennial herbs with a taproot, usually not aromatic; stems procumbent to erect, usually branched; leaves basal and cauline, but all similar except the basal ones petiolate; heads usually radiate, borne singly or in corymbs; involucre hemispheric; phyllaries in 2–5 series, unequal or nearly equal, usually glabrous, the margins scarious; receptacle epaleate; ray flowers white, pistillate, fertile, rarely absent; disc flowers numerous, tubular, yellow, bisexual, fertile, 5-lobed; cypselae trigonous, conspicuously 3- or 5-ribbed, glabrous; pappus a crown of minute scales or absent.

Approximately 40 species are in the genus. One is adventive in the United States.

1. **Tripleurospermum inodorum** (L.) Schultz-Bip. Tanaceteen. 32. 1844. Fig.127.
Matricaria inodora L. Fl. Suec., ed. 2, 297. 1755.
Matricaria perforata Merat, Nouv. Fl. Env. Paris 332. 1812.

Non-aromatic annual herb from a taproot; stems ascending to erect, much branched, glabrate, to 75 cm tall; leaves alternate, 2- to 3-pinnate, the divisions filiform, glabrous; heads radiate, few to numerous, to 3 cm across, borne singly or in corymbs; involucre hemispheric; phyllaries in 3 series, narrowly oblong, obtuse at the apex, often brownish, with scarious margins; receptacle epaleate; ray flowers up to 30, to 2 cm long, white, pistillate, fertile; disc flowers numerous, tubular, bisexual, yellow, 5-lobed, 1.0–3.5 mm high; cypselae trigonous, 3-ribbed, glabrous, gland-dotted, pale brown, to 1 mm long; pappus a crown of minute scales.

Common Name: Scentless chamomile; mayweed.
Habitat: Disturbed soil.
Range: Introduced into most of the United States and Canada.
Illinois Distribution: Boone, DuPage, Grundy, Kane, and Lake counties.

In the past, this species has been included in *Matricaria*, either as *M. inodora* or *M. perforata*. It differs from *Matricaria* by its lack of an odor.

The flowers bloom from June to September.

127. *Tripleurospermum inodorum* (Scentless chamomile).

a. Upper part of plant.
b. Leaf.
c. Flowering head.
d. Involucre.
e. Phyllary.
f. Ray flower.
g. Disc flower.
h. Cypsela.

29. **Leucanthemum** Mill.—Daisy

Perennial herbs from rhizomes; leaves basal or basal and cauline; heads radiate, in groups of 1–3; involucre hemispheric; phyllaries in 3 or 4 series, unequal, scarious along the margins; receptacle epaleate; ray flowers up to 40, usually white, pistillate, fertile; disc flowers numerous, tubular, yellow, bisexual, fertile, 5-lobed; cypselae 10-ribbed, glabrous; pappus absent.

As many as 40 species may be in this genus, all native to Europe. One is adventive in the United States.

1. **Leucanthemum vulgare** Lam. Fl. Franc. 2:137. 1779. Fig. 128.
Chrysanthemum leucanthemum L. Sp. Pl. 2:888. 1753.

Perennial herb from rhizomes; stems erect, branched or unbranched, glabrous or puberulent, to 1 m tall; leaves of two types: basal leaves obovate to spatulate, dentate or pinnately 3- to 7-lobed, glabrous or nearly so, long-petiolate, to 4 cm long, to 2 cm wide; cauline leaves alternate, linear to narrowly spatulate, acute at the apex, tapering to the sessile and sometimes clasping base, dentate or nearly entire, to 8 cm long, to 1.5 cm wide; heads radiate, 1–3, to 3.5 cm across; involucre hemispheric, 8–12 mm high; phyllaries in 3 or 4 series, oblong to lanceolate, obtuse at the apex, glabrous or nearly so; receptacle epaleate; ray flowers up to 35, 12–30 mm long, white, pistillate, minutely notched at the tip; disc flowers numerous, tubular, yellow, bisexual; cypselae of ray flowers angular, 10-ribbed, glabrous, 1.5–2.5 mm long; pappus absent.

Common Name: Ox-eye daisy.
Habitat: Old fields, pastures, roadsides, edge of woods, disturbed soil.
Range: Native to Europe; adventive in much of the United States.
Illinois Distribution: Common throughout the state; probably in every county.

This handsome species is recognized by its pinnatifid leaves and its large flowering heads.

For many years it was known as *Chrysanthemum leucanthemum*.

This species flowers from May through August.

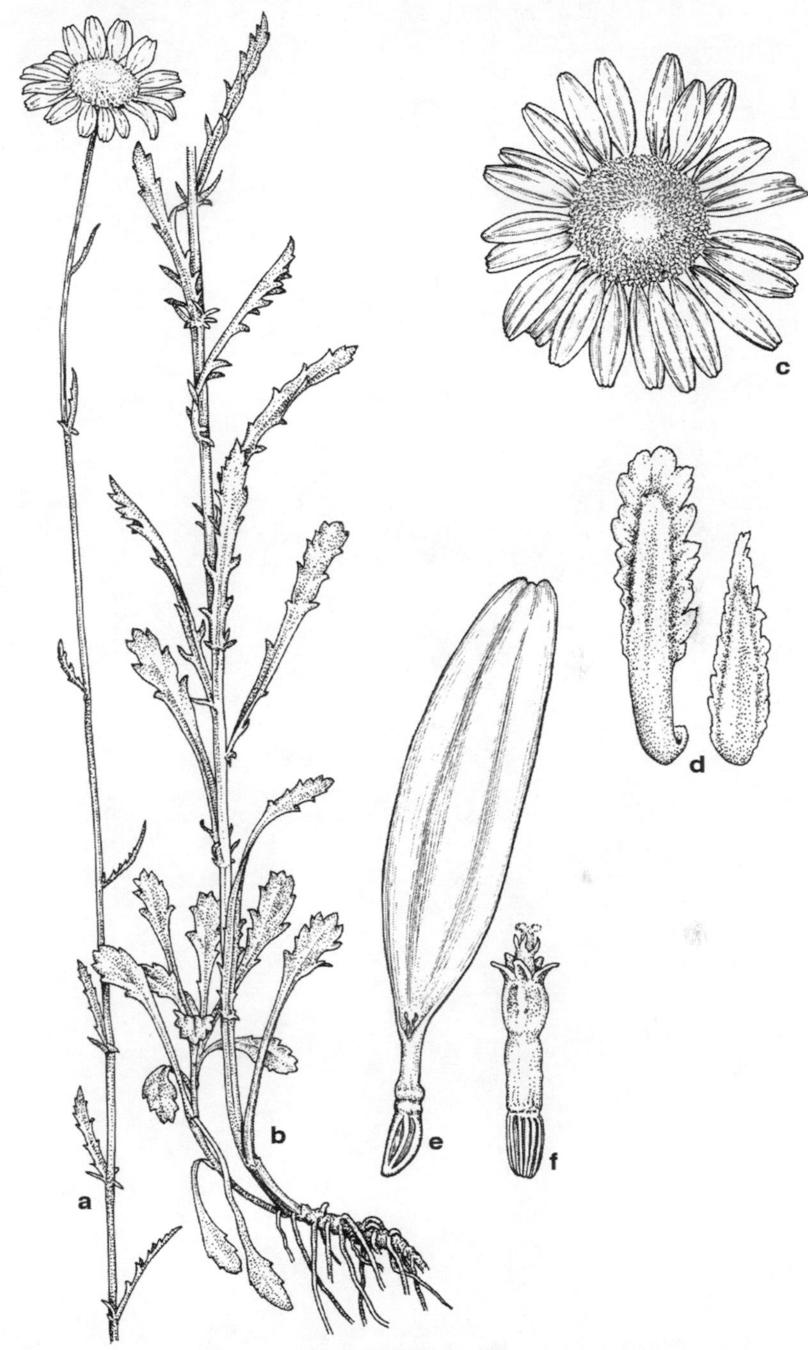

128. *Leucanthemum vulgare* (Ox-eye daisy).

a. Upper part of plant.
b. Lower part of plant.
c. Flowering head.

d. Phyllaries.
e. Ray flower.
f. Disc flower.

Excluded Species

Glossary

Literature Cited

Index of Scientific Names

Index of Common Names

Excluded Species

Artemisia canadensis Michx. Babcock (1872), Higley and Raddin (1891), and Pepoon (1927) erroneously used this binomial for *Artemisia caudata* Michx.

Artemisia dracunculus L. Cronquist (1952) called *Artemisia dracunculoides* Pursh by this binomial.

Artemisia glauca Pallas ex Willd. Fernald (1950) used this name for *Artemisia dracunculoides* Pursh.

Artemisia longifolia Nutt. Pepoon (1927) used this name erroneously for *Artemisia ludoviciana* Nutt.

Artemisia procera Willd. Pepoon (1927) called *Artemisia abrotanum* L. by this name.

Aster angustus Nees. Higley and Raddin (1891) and Pepoon (1927) used this name for *Brachyactis ciliata* (Ledeb.) Ledeb.

Aster carneus Nees. Engelmann (1843), Short (1845), Mead (1846), Lapham (1857), Patterson (1876), and Schneck (1876) have reported this binomial for specimens in Illinois. The plants are actually *Symphyotrichum praealtum* (Poir. in Lam.) G.L. Nesom.

Aster concolor L. Short (1845) and Lapham (1857) reported this plant from Illinois. It actually refers to *Symphyotrichum sericeum* (Vent.) G.L. Nesom.

Aster corymbosus Ait. This is an error for *Eurybia furcata*, although it was reported from Illinois by Vasey (1860, 1861), Patterson (1874, 1876), Brendel (1870), and Huett (1891).

Aster gracilis Nutt. Short (1845) and Lapham (1857) erroneously called *Symphyotrichum turbinellus* by this name.

Aster infirmus Michx. Brendel's report (1887) of this species in Illinois was an error for *Doellingeria umbellata*.

Aster junceus Ait. Brendel (1887), Fuller (1946), and Fuller, Fell, and Fell (1949) used this name for *Symphyotrichum boreale*.

Aster longifolius Lam. Although this binomial has been attributed to plants in Illinois by Babcock (1873), Patterson (1876), Huett (1897), Pepoon (1927), Jones (1950), Jones and Fuller (1955), and Jones (1963), these references were actually misidentifications for *Symphyotrichum boreale*.

Aster missouriensis Britt. in Britt. & Brown, *non* (Nutt.) Kuntze. Wiegand (1928) used this name for *Symphyotrichum ontarionis.*

Aster novi-belgii L. I erroneously attributed this Coastal Plain species to Illinois based upon my identification of a few specimens from Cook, Lake, and McHenry counties. I believe these specimens are actually *Symphyotrichum boreale* that have slightly wider leaves and slightly longer involucres than typical specimens of *S. boreale.*

Aster paniculatus Lam. Up until 1952, Illinois botanists used this binomial erroneously for *Symphyotrichum lanceolatum.*

Aster salicifolius Lam. Botanists in Illinois before 1952 used this name erroneously for *Symphyotrichum praealtum.*

Aster tenuifolius L. Mead (1836) and Patterson (1876) used this binomial erroneously for *Symphyotrichum parviceps.*

Aster tradescanti Lam. Many Illinois botanists up until 1952 used the binomial of this species of the northeastern United States erroneously for what is now considered to be *Symphyotrichum lanceolatum* var. *interior.*

Boltonia glastifolia (Hill) L'Her. Although several early Illinois botanists used this binomial for *Boltonia asteroides var. recognita, Boltonia glastifolia* belongs to a different species.

Chrysopsis baillardii Rydb. Fernald (1950) attributed this western plant to Illinois. It was in error for *Chrysopsis camporum.*

Chrysopsis mariana (L.) Ell. This species was attributed to Illinois by Short (1845), Mead (1846), Lapham (1857), and McDougall (1936), an error for *Chrysopsis camporum.*

Erigeron quercifolius Lam. Mead (1836) used this binomial for *Erigeron philadelphicus.* It is a different species from the southeastern United States.

Euthamia tenuifolia (Pursh) Greene. This species was erroneously reported from Illinois by Britton and Brown (1913).

Solidago gilmanii (Gray) Steele. Although Patterson (1876) cited this plant as being "near Chicago" and Higley and Raddin (1891) observed it "in sandy soil near the lake shore, chiefly southward, infrequent and rare," I have found no evidence that this plant was ever in Illinois. However, it may well be found on the Illinois side of Lake Michigan.

Solidago houghtonii Torr. & Gray. Although Friesner (1933) and Fernald (1950) attributed this species to Illinois, I have seen no evidence that it occurs in the state.

Solidago macrophylla Pursh. Snare and Hicks (1898) reported this plant from Illinois, but it was based on specimens of *S. flexicaulis*.

Solidago missouriensis Nutt. Although I have attributed this species to Illinois in the past, the specimens I observed were actually *S. juncea*.

Solidago moritura Steele. This is a synonym for *S. glaberrima*. The type is from Urbana, Champaign County, collected by E. S. Steele on August 10, 1910.

Solidago odora Ait. Patterson (1876) and Schneck (1876) attributed this plant to Illinois. It may well be in the state since it occurs in Kentucky and Missouri, but there are apparently no specimens of it having been found in Illinois.

Solidago serotina Ait. Until 1950, most Illinois botanists called *S. gigantea* by this name. *Solidago serotina* is a different species not known from Illinois.

Solidago squarrosa Muhl. Ries (1939) reported this plant from Illinois, but the report actually refers to *S. speciosa*.

Solidago stricta Ait. Although Bebb (1859), Vasey (1861), Babcock (1872), Patterson (1876), and Brendel (1857) attributed this southeastern United States species to Illinois, the specimens were actually of *S. uliginosa*.

Solidago tenuifolia Pursh. Several botanists have used this name for our *Euthamia graminifolia*. *Solidago tenuifolia* is a different species not found in Illinois.

Solidago uniligulata (DC.) Porter. Some botanists used this binomial for *S. uliginosa*.

Glossary

acuminate. Gradually tapering to a long point.
acute. Sharply tapering to a point.
annual. A plant that lives for only one growing season.
anthesis. Flowering time.
apiculate. Abruptly short-pointed at the tip.
appressed. Lying flat against the surface.
aristate. Bearing an awn.
array. An arrangement of flowering heads in an inflorescence in the Asteraceae.
attenuate. Gradually becoming narrowed.
auriculate. Bearing an earlobe process.
awn. A bristle usually terminating a structure.
axil. The junction of two structures.
axillary. Borne in an axil.

barbellate. Bearing barbs or setae with downward pointing prickles.
beaked. Having a short point at the tip.
biennial. A plant that completes its life cycle in two years and then perishes.
bisexual. Said of a flower bearing both staminate and pistillate parts.
bract. An accessory structure on the peduncle bearing flowering heads.
bracteate. Bearing one or more bracts.
bracteolate. Bearing one or more bracteoles.
bracteole. A secondary bract.
bractlet. A small bract.
bristle. A stiff hair or hairlike growth; a seta.

caducous. Falling away early.
campanulate. Bell-shaped.
canescent. Grayish-hairy.
capillary. Threadlike.
capitate. Forming a head.
capitulum. A flowering head in the Asteraceae.
caudex. The woody base of a perennial plant.
cauline. Belonging to a stem.
cespitose. Growing in tufts.
chaff. A scale or group of scales.
chartaceous. Papery in texture.
cilia. Marginal hairs.
ciliate. Bearing cilia.
ciliolate. Bearing small cilia.
clasping. Said of a leaf whose base wraps partway around the stem.
columnar. Shaped like a column or cylindrical upright structure.
compressed. Flattened.
concave. Curved on the inner surface; opposed to convex.
convex. Curved on the outer surface; opposed to concave.
cordate. Heart-shaped.
coriaceous. Leathery.
corolla. That part of a flower composed of petals.
corymb. A type of inflorescence where the pedicellate flowers are arranged along an elongated axis but with the flowers all attaining about the same height.
corymbiform. Shaped like a corymb.
crenate. With round teeth.
cuspidate. Terminating in a very short point.
cyme. A type of broad and flattened inflorescence in which the central flowers bloom first.
cymose. Bearing a cyme.
cypsela. The fruit in the Asteraceae.

deciduous. Falling away.
decumbent. Lying on the ground, but with the extremities ascending.
decurrent. Adnate to the petiole or stem and then extending beyond the point of attachment.
dentate. With sharp teeth, the tips of which project outward.
denticulate. With small, sharp teeth, the tips of which project outward.
diffuse. Loosely spreading.
disc. The group of flowers that make up the center of a flowering head in the Asteraceae.
disciform. Bearing only disc flowers, but the peripheral flowers are filiform and usually pistillate.
discoid. Bearing only disc flowers.

eciliate. Without cilia.
eglandular. Without glands.
ellipsoid. Referring to a solid object that is broadest at the middle, gradually tapering to both ends.
elliptic. Broadest at the middle, gradually tapering to both ends.
entire. Without any projections along the edge.
epaleate. Without palea or outgrowths from the receptacle in the Asteraceae.
epunctate. Without dots.
erose. With an irregularly notched tip.

fascicle. Cluster.
fibrous. Referring to roots borne in tufts.
filiform. Threadlike.
flexuous. Zigzag.
fusiform. Spindle-shaped.

glabrate. Becoming smooth.
glabrous. Without pubescence or hairs.
gland. An enlarged, usually spherical body functioning as a secretory organ.
glandular. Bearing glands.
glaucous. With a whitish covering that may be rubbed off.
globose. Round; globular.
glomerule. A small, compact cluster.
glutinous. Covered with a sticky secretion.

hirsute. Bearing stiff hairs.
hirtellous. Finely hirsute.
hispid. Bearing rigid hairs.
hyaline. Transparent.

inferior. Referring to the position of the ovary when it is surrounded by the adnate portion of the floral tube or is embedded in the receptacle.
inflorescence. A cluster of flowers.
involucre. The collection of phyllaries around a flower cluster in the Asteraceae.
involute. Rolled inward.

keel. A ridgelike process.

laciniate. Divided into narrow, pointed divisions.
lamina. A blade.
lanceolate. Lance-shaped; broadest near the base, gradually tapering to the narrower apex.
lanceoloid. Referring to a solid object that is broadest near the base, gradually tapering to the narrower apex.
latex. Milky juice.
leaflet. An individual unit of a compound leaf.
ligulate. Flowers with ligules.
ligule. A flat, narrow petal-like structure of a flower in the Asteraceae; a ray.
linear. Elongated and narrow in width throughout.
lustrous. Shiny.

mucro. A short, abrupt tip.
mucronate. Possessing a short, abrupt tip.
mucronulate. Possessing a very short, abrupt tip.

obconic. Reverse cone-shaped.
oblanceolate. Reverse lance-shaped; broadest at apex, gradually tapering to narrow base.
oblong. Broadest at the middle and tapering to both ends, but broader than elliptic.
oblongoid. Referring to a solid object that is broadest at the middle and tapering to both ends.
obovate. Broadly rounded at the apex, becoming narrowed below.
obovoid. Referring to a solid object that is broadly rounded at the apex, becoming narrowed below.
obtuse. Rounded at the apex.
orbicular. Round.
oval. Broadly elliptic.
ovate. Broadly rounded at base, becoming narrowed above; broader than lanceolate.
ovoid. Referring to a solid object that is broadly rounded at base, becoming narrowed above.

palea. An outgrowth from the receptacle in the Asteraceae.
paleate. Bearing paleae.
palmate. Divided radiately, like the fingers from a hand.
panicle. A type of inflorescence composed of several racemes.
paniculate. Having the flowers in a panicle.

pannose. Having the texture of felt.
pappus. Various kinds of structures attached to the cypsela in the Asteraceae.
pedicel. The stalk of a flower.
pedicellate. Said of a flower that has a pedicel.
peduncle. The stalk of an inflorescence.
perennial. Living more than two years.
perfect. Said of a flower that has both stamens and pistils.
petiolate. Having a petiole.
petiole. The stalk of a leaf.
phyllary. A bract subtending a flowering head in the Asteraceae.
pilose. Bearing soft hairs.
pinna. A primary division of a compound blade.
pinnate. Divided once into distinct segments.
pinnatifid. Said of a simple leaf or leaf part that is cleft or lobed only partway to its axis.
pinnatisect. Divided in a pinnate manner.
pistillate. Bearing pistils but not stamens.
plumose. Bearing fine hairs, like the plume of a feather.
prostrate. Lying flat.
puberulent. Bearing minute hairs.
pubescent. Bearing some kind of hairs.
punctate. Dotted.
pustulate. Having small, pimplelike swellings.
pyramidal. Shaped like a pyramid.

raceme. A type of inflorescence where pedicellate flowers are arranged along an elongated axis.
radiate. Bearing ray flowers in the Asteraceae.
ray. A flat flower in the Asteraceae; a ligule.
receptacle. A structure to which all flowers are attached in the capitulum in the Asteraceae.
reflexed. Turned downward.
resinous. Producing a sticky secretion, or resin.
reticulate. Resembling a network.
rhizomatous. Bearing rhizomes.
rhizome. An underground horizontal stem bearing nodes, buds, and roots.
rosette. A cluster of leaves in a circular arrangement at the base of a plant.
rugose. Wrinkled.
rugulose. With small wrinkles.

sagittate. Shaped like an arrowhead.
scabrellous. Slightly rough to the touch.
scabrous. Rough to the touch.
scarious. Thin and membranous.
secund. Borne on one side.
septate. With dividing walls.
sericeous. Silky; bearing soft, appressed hairs.
serrate. With teeth that project forward.
serrulate. With very small teeth that project forward.
sessile. Without a stalk.
seta. Bristle.
setaceous. Bearing bristles or setae.
setose. Bearing setae.
spatulate. Oblong but with the basal end elongated.
spicate. Bearing a spike.
spike. A type of inflorescence where sessile flowers are arranged along an elongated axis.
spinescent. Becoming spiny.
spinose. Bearing spines.
spinulose. Bearing small spines.
squarrose. With tips spreading or recurved.
staminate. Bearing stamens but not pistils.
stipe. A stalk.
stipitate. Bearing a stalk.
stolon. A slender, horizontal stem on the surface of the ground.
stoloniferous. Bearing stolons.
stramineous. Straw-colored.
striate. Marked with grooves.
strigillose. Bearing short, appressed, straight hairs.
strigose. Bearing appressed, straight hairs.
subacute. Nearly tapering to a short point.
subcordate. Nearly cordate.
suborbicular. Nearly round.
subulate. With a very short, narrow point.
succulent. Fleshy.

terete. Round in cross-section.
ternate. Divided three times.
thyrse. A mixed inflorescence containing panicles and cymes.
thyrsoid. Bearing a thyrse.

trigonous. Triangular in cross-section.
truncate. Abruptly cut across.
turbinate. Top-shaped; shaped like a turban.

undulate. Wavy.
urceolate. Urn-shaped.

villous. Bearing long, soft, slender, unmatted hairs.
viscid. Sticky.

whorl. An arrangement of three or more structures at a point.

Literature Cited

Babcock, H. H. 1872. The flora of Chicago and vicinity. *Lens* 2:65–71.

———. 1873. The flora of Chicago and vicinity. *Lens* 2:248–50.

Bebb, M. S. 1859. List of plants occurring in the northern counties of the state of Illinois. *Transactions of the Illinois State Agricultural Society* 3:586–87.

Brendel, F. 1870. Distribution of immigrant plants. *American Entomologist and Botanist* 2:378–79.

———. 1887. *Flora Peoriana*. Peoria, Illinois.

Britton, N. L., and A. Brown. 1913. *An illustrated flora of the northern United States and Canada*. 2nd ed. New York: New York Botanical Garden.

Cronquist, A., in H. A. Gleason. 1952. *Illustrated Flora of the Northeastern United States*. New York: New York Botanical Garden.

Engelmann, G. 1843. Catalogue of collections of plants made in Illinois and Missouri by Charles A. Geyer. *American Journal of Science* 46:94–104.

Fernald, M. L. 1950. *Gray's Manual of Botany*. 8th ed. 1632 pp. New York: American Book Company.

Flora of North America. Vol. 20, *Magnoliophyta: Asteridae (in part): Asteraceae, Part 2*. 2006. Oxford: Oxford University Press.

Friesner, R. C. 1933. The genus *Solidago* in northeastern North America. *Butler University Botanical Studies* 3:1–64.

Fuller, G. D. 1946. A check list of the vascular plants of Jo Daviess County, Illinois. *Transactions of the Illinois Academy of Science* 38:51–63.

Fuller, G. D., E. W. Fell, & G. B. Fell. 1949. Checklist of the vascular plants of Winnebago County, Illinois. *Transactions of the Illinois Academy of Science* 42:68–79.

Higley, W. K., & C. S. Raddin. 1891. Flora of Cook County, Illinois, and a part of Lake County, Indiana. *Bulletin of the Chicago Academy of Science* 2:1–168.

Huett, J. W. 1897. Essay toward a natural history of LaSalle County, Illinois. *Flora LaSallensis*, part 1. LaSalle, Ill.: published by author.

Jones, G. N. 1950. *Flora of Illinois*. 2nd ed. 368 pp. American Midland Naturalist Monograph Number 5. Notre Dame, Ind.: University of Notre Dame Press.

———. 1963. *Flora of Illinois*. 3rd ed. 402 pp. Notre Dame, Ind.: University of Notre Dame Press.

Jones, G. N., & G. D. Fuller. 1955. *Vascular plants of Illinois*. 593 pp. Urbana: University of Illinois Press; Springfield: Illinois State Museum.

Lapham, I. A. 1857. Catalogue of the plants of the state of Illinois. *Transactions of the Illinois State Agricultural Society* 2:492–550.

McDougall, W. B. 1936. *Fieldbook of Illinois wild flowers*. 406 pp. Urbana: Illinois Natural History Survey.

Mead, S. B. 1846. Catalogue of plants growing spontaneously in the state of Illinois. *Prairie Farmer* 6:35–36; 60; 93; 119–22.

Mohlenbrock, R. H. 1975. *Guide to the vascular flora of Illinois*. 494 pp. Carbondale: Southern Illinois University Press.

———. 1986. *Guide to the vascular flora of Illinois*. Revised ed. 507 pp. Carbondale: Southern Illinois University Press.

———. 2002. *Vascular flora of Illinois*. 491 pp. Carbondale: Southern Illinois University Press.

Patterson, H. N. 1874. *A list of plants collected in the vicinity of Oquawka, Henderson County*. 18 pp. Oquawka, Ill.: published by the author.

———. 1876. *Catalogue of the phaenogamous and vascular cryptogamous plants of Illinois*. 54 pp. Oquawka, Ill.: published by the author.

Pepoon, H. S. 1927. An annotated flora of the Chicago region. *Bulletin of the Chicago Academy of Science* 8:1–554.

Schneck, J. 1876. Catalogue of the flora of the Wabash Valley. *Annual Report of the Geological Survey of Indiana* 7:504–79.

Short, C. W. 1845. Observations on the botany of Illinois. *Western Journal of Medicine and Surgery* 3:185–98.

Snare, W., & E. W. Hicks. 1898. *Check list of plants in the Boardman Collection, Toulon Academy*. 29 pp. Published by the authors.

Vasey, G. 1860. Additions to the Illinois flora. *Prairie Farmer* 22:119.

———. 1861. Additions to the flora of Illinois. *Transactions of the Illinois Natural History Society* 1:139–43.

Wiegand, K. M. 1933. *Aster paniculatus* and some of its relatives. *Rhodora* 35:16–38.

Index of Scientific Names

Names in roman type are accepted names, while those in italics are synonyms and are not considered valid. Numbers in boldface refer to pages that have illustrations.

Index of Common Names

Robert H. Mohlenbrock taught botany at Southern Illinois University Carbondale for thirty-four years. Since his retirement in 1990, he has served as a senior scientist for Biotic Consultants, teaching wetland identification classes around the country. Among his more than sixty books are *Vascular Flora of Illinois* and *Field Guide to the U.S. National Forests*.